T0236245

Research Advances in ADHD and Technology

Synthesis Lectures on Assistive, Rehabilitative, and Health-Preserving Technologies

Editors
Ronald M. Baecker, *University of Toronto*
Andrew Sixsmith, *Simon Fraser University*

Advances in medicine allow us to live longer, despite the assaults on our bodies from war, environmental damage, and natural disasters. The result is that many of us survive for years or decades with increasing difficulties in tasks such as seeing, hearing, moving, planning, remembering, and communicating.

This series provides current state-of-the-art overviews of key topics in the burgeoning field of assistive technologies. We take a broad view of this field, giving attention not only to prosthetics that compensate for impaired capabilities, but to methods for rehabilitating or restoring function, as well as protective interventions that enable individuals to be healthy for longer periods of time throughout the lifespan. Our emphasis is in the role of information and communications technologies in prosthetics, rehabilitation, and disease prevention.

Research Advances in ADHD and Technology
Franceli L. Cibrian, Gillian R. Hayes, and Kimberley D. Lakes

AgeTech, Cognitive Health, and Dementia
Andrew Sixsmith, Judith Sixsmith, Mei Lan Fang, and Becky Horst

Interactive Technologies and Autism, Second Edition
Julie A. Kientz, Gillian R. Hayes, Matthew S. Goodwin, Mirko Gelsomini, and Gregory D. Abowd

Zero-Effort Technologies: Considerations, Challenges, and Use in Health, Wellness, and Rehabilitation, Second Edition
Jennifer Boger, Victoria Young, Jesse Hoey, Tizneem Jiancaro, and Alex Mihailidis

Human Factors in Healthcare: A Field Guide to Continuous Improvement
Avi Parush, Debi Parush, and Roy Ilan

© Springer Nature Switzerland AG 202, corrected publication 2023
Reprint of original edition © Morgan & Claypool 2021

All rights reserved. No part of this publication may be reproduced, stored in a retrieval system, or transmitted in any form or by any means—electronic, mechanical, photocopy, recording, or any other except for brief quotations in printed reviews, without the prior permission of the publisher.

Research Advances in ADHD and Technology
Franceli L. Cibrian, Gillian R. Hayes, and Kimberley D. Lakes

ISBN: 978-3-031-00478-0   print
ISBN: 978-3-031-01606-6   ebook
ISBN: 978-3-031-00040-9   hardcover

DOI 10.1007/978-3-031-01606-6

A Publication in the Springer series
SYNTHESIS LECTURES ON ASSISTIVE, REHABILITATIVE, AND HEALTH-PRESERVING TECHNOLOGIES
Lecture #15
Series Editors: Ron Baecker, University of Toronto and Andrew Sixsmith, Simon Fraser University and AGE-WELL NCE

Series ISSN 2162-7258 Print 2162-7266 Electronic

Research Advances in ADHD and Technology

Franceli L. Cibrian
Chapman University

Gillian R. Hayes
University of California, Irvine

Kimberley D. Lakes
University of California, Riverside

SYNTHESIS LECTURES ON ASSISTIVE, REHABILITATIVE, AND HEALTH-PRESERVING TECHNOLOGIES #15

ABSTRACT

Attention Deficit Hyperactivity Disorder (ADHD) is the most prevalent childhood psychiatric condition, with estimates of more than 5% of children affected worldwide, and has a profound public health, personal, and family impact. At the same time, a multitude of adults, both diagnosed and undiagnosed, are living, coping, and thriving while experiencing ADHD. It can cost families raising a child with ADHD as much as five times the amount of raising a child without ADHD (Zhao et al. 2019). Given the chronic and pervasive challenges associated with ADHD, innovative approaches for supporting children, adolescents, and adults have been engaged, including the use of both novel and off-the-shelf technologies. A wide variety of connected and interactive technologies can enable new and different types of sociality, education, and work, support a variety of clinical and educational interventions, and allow for the possibility of educating the general population on issues of inclusion and varying models of disability.

This book provides a comprehensive review of the historical and state-of-the-art use of technology by and for individuals with ADHD. Taking both a critical and constructive lens to this work, the book notes where great strides have been made and where there are still open questions and considerations for future work. This book provides background and lays foundation for a general understanding of both ADHD and innovative technologies in this space. The authors encourage students, researchers, and practitioners, both with and without ADHD diagnoses, to engage with this work, build upon it, and push the field further.

KEYWORDS

Attention Deficit Hyperactivity Disorder, ADHD, human-computer interaction, cognition, social interaction, social skills, education, disability, human development, interactive technologies, user experience, mobile computing, shared active surfaces, tabletop computing, virtual reality, multi-sensory environments, augmented reality, sensors, wearable computing, robots, robotics, natural user interfaces, natural input, tangible computing, tactile computing, eye tracking, behavioral intervention

Franceli L. Cibrian:
To Armando, Rome, and Francisco

Gillian R. Hayes:
To Steve, Warner, and William

Kimberley D. Lakes:
To Dr. Francis Crinella and James, Emma, Micah, and Elijah

Contents

XV

Acknowledgments

We would like to thank the following people who provided input directly on the writing of this manuscript, providing reviews, discussions, images, or comments on the content and writing: Sabrina E.B. Schuck, Mirko Gelsomini, Jeffrey Krichmar, Patrizia Marti, and Jesus A. Beltran. Additionally, Latoya Wilson has been an enormous help in organizing time and deadlines to ensure this book actually got written. Finally, Gabriela Marcu and Paul Marshall provided extensive comments and edits that improved the final version.

We wish to thank the following people for their influence on our thinking and informing our understanding of the fields of ADHD, human-computer interaction, and health informatics in significant ways: Sabrina E. B. Schuck, LouAnne Boyd, Julie Kientz, Nancy Donnelly, and the children, parents, and teachers at The Children's School in Irvine, California.

Finally, we wish to acknowledge the following funding sources that supported the authors during the writing of this manuscript: National Institutes of Health (NIH), the Agency for Healthcare Research and Quality (AHRQ) under award number 1R21HS026058, and the Jacobs Foundation Advanced Research Fellowship. The content is solely the responsibility of the authors and does not necessarily represent the official views of the NIH, AHRQ, nor the Jacobs Foundation.

The original version of this book was revised: Author affiliation country is incorrect for Gillian R. Hayes and Kimberley D. Lakes. The affiliation country should be USA not UK for authors Gillian R. Hayes and Kimberley D. Lakes. This has now been corrected. The updated online version of this book can be found at DOI 10.1007/978-3-031-01607-3_11.

CHAPTER 1

Introduction

Attention Deficit Hyperactivity Disorder (ADHD) is the most prevalent childhood psychiatric condition, with worldwide prevalence estimates of 5% and 7.2% (Polanczyk, et al., 2014; Thomas et al., 2015), and profound public health, personal, and family impacts. In 2013, in a European setting estimate that the average total ADHD-related costs ranged from €9,860 to €14,483 per patient and annual national costs were between €1,041 and €1,529 million (M) (Le et al., 2014). In the U.S. in 2000, a study estimated that the excess cost of ADHD was $31.6 billion (1.6 billion was for the treatment; $14.2 billion for healthcare cost and $3.7 billion was for the work loss) (Birnbaum et al., 2005; Doshi et al., 2012). In the Republic of Korea, the total economic burden of ADHD was US$47.55 million, which accounted for approximately 0.004% of Korean gross domestic product in 2012 (Hong et al., 2020). Given the international differences in the medical care system, it is difficult to generalize a global cost and there is not a lot of data about the burdensome costs from countries of the Global South; still, the burden may be similar or greater, although it may not be recognized to the same extent.

At the same time, a multitude of adults are living, coping, and thriving while experiencing ADHD, both diagnosed and undiagnosed. Given the chronic and pervasive challenges associated with ADHD, innovative approaches for supporting people with ADHD across the lifespan have been engaged, including the use of both novel and off-the-shelf technologies. A wide variety of connected and interactive technologies can enable new and different types of sociality, education, and work, support a variety of clinical and educational interventions, and allow for the possibility of educating the general population on issues of inclusion and varying models of disability. These technologies are used for traditional mental health care, such as diagnosis and assessment, as well as addressing the primary mental health concerns of ADHD, including cognition and attention. We intentionally engage a variety of technologies beyond this, however. Notably, we explore behaviors that can put people with ADHD in conflict with those without ADHD and discuss how technologies can support self-regulation and management of behavior to fit cultural and societal norms and social engagement with those without ADHD. We also explore technologies most concerned with promoting the pragmatic outcomes of academic, work, and daily lives. Our goals are to describe both a broad base of existing and emerging technological approaches as well as the large scope of potential opportunities and challenges in the space of ADHD that future technologies could address.

Since 2004, research papers describing potential applications of technology to ADHD have increased dramatically; the ACM Digital Library (2019, https://dl.acm.org/) indicated that from 2004–2019 the numbers of publications focused on ADHD and technology increased exponentially, with the fastest growth occurring in the last ten years. As scientific interest has grown, commercial interest has developed even more rapidly. Many commercial products are currently available on the market, advertising clinical benefits to individuals with ADHD and their families. The efficacy of these products, however, is largely unknown.

Limited reviews of the scientific literature surrounding ADHD and technology use do exist and are included in this book. Every day, people with ADHD and their families, co-workers, friends, and teachers are adapting and adopting interactive technologies in novel ways not cataloged by any researchers. Thus, this review does not attempt to make clinical recommendations nor provide a comprehensive examination of all technology use surrounding ADHD. Rather, this volume focuses on describing current research—technological, social-behavioral, and medical—in the broad spaces of ADHD assessment, diagnosis, treatment, and self-determination.

Each of the authors of this book has worked with and/or taught people with ADHD of all ages for years. We come to this work as researchers, clinicians, educators, family members, and friends of people with ADHD. Additionally, we have sought and incorporated feedback from people with ADHD, including both children and adults, throughout our time working on this manuscript and are grateful for their lived experiences, which have informed this book. Our intention with this book is to help readers understand ADHD as both a medical and sociological construct and provide a broad review of the research literature that demonstrates the role technology has served to this point. Finally, we hope through this work to identify some gaps, biases, and challenges that we as a community have yet to overcome.

This book provides an overview of the historical and state-of-the-art use of technology by and for individuals with ADHD with deep dives into a subset of research to further illustrate the trends we see in the space. We take both a critical and constructive lens to this work, noting where we have made great strides and where there are still open questions and considerations that must be engaged. This book provides background and lays foundation for a general understanding of both ADHD and innovative technologies in this space. We encourage students, researchers, and practitioners, both with and without ADHD diagnoses, to engage with this work, build upon it, and push the field ahead in order to support the needs of children, adolescents, and adults with ADHD.

1.1   THE EVOLUTION OF ADHD AS A DIAGNOSTIC CATEGORY

The historic 1980 revision of the *Diagnostic and Statistical Manual of the American Psychiatric Association* (DSM-III, 1980) introduced the diagnostic categories of Attention Deficit Disorder without (ADD) or with Hyperactivity (ADDH). These categories replaced the prior diagnostic category of hyperkinetic reaction of childhood (DSM-II, 1978). The DSM-III diagnostic criteria described symptoms of inattention, impulsivity, and hyperactivity. Subsequent revisions to the manual (e.g., DSM-IIIR, DSM-IV 1994, DSM-5) expanded on these categories and symptoms. Today, the DSM-5 diagnosis of ADHD falls within the broader grouping of Neurodevelopmental Disorders, a grouping of conditions that emerge during early development. DSM-5 diagnostic criteria are grouped as either inattention or hyperactivity-impulsivity symptoms. The earlier term "ADD" is no longer applied; rather, the DSM-5 uses the diagnosis of ADHD, which can be specified as "Combined presentation" (six or more symptoms of inattention and six or more symptoms of hyperactivity-impulsivity have been present for at least six months), "Predominantly inattentive presentation" (sufficient inattention symptoms are present but hyperactivity-impulsivity symptoms do not meet the threshold), or "Predominantly hyperactive/impulsive presentation" (symptoms of hyperactivity-impulsivity meet the threshold, but symptoms of inattention do not).

The DSM-5 provides examples of how inattention and hyperactivity-impulsivity impair daily functioning. Symptoms of inattention can include having difficulty focusing on a task, failing to attend to details, demonstrating disorganization, lacking persistence, losing things, and forgetting to do things. Symptoms of hyperactivity often include fidgeting, having difficulty remaining in place, and extreme restlessness. Impulsivity can be exhibited by acting without thinking, talking excessively, interrupting others, and having difficulty waiting in line. The DSM-5 requires that these difficulties interfere with functioning to meet the threshold for diagnosis and that alternative explanations for these difficulties are considered and ruled out.

Researchers (e.g., Swanson, Wigal, and Lakes, 2009) have described attention and behavioral control (the ability to regulate behavior and control one's impulses) as occurring across a spectrum, noting that individuals with ADHD fall at an "extreme" on the spectrum and exhibit impairment in daily functioning. In other words, attention and self-regulation are exhibited to some degree in all persons. Some demonstrate very strong abilities to pay attention and sustain attention over a long period of time, while others struggle to do so. Likewise, some individuals display impressive levels of self-control, while others appear to have very little self-control and frequently display impulsivity. These cognitive and behavioral skills have been described as critical to success in life. In a rigorous longitudinal study, Moffit et al. (2011) followed nearly 1,000 children from birth into adulthood and concluded that self-control predicted physical health, substance dependence, personal finances, and criminality, producing a gradient that indicated that as self-control improved, corresponding outcomes improved as well. Their study continues to have important implications for society at large, as well as individuals with ADHD who struggle with self-control: improving self-control can improve outcomes across a wide range of domains (health, education, wealth, and citizenship). A large body of evidence in the field of ADHD research similarly demonstrated that difficulties with self-control are associated with poor educational, occupational, and social outcomes. Collectively, this research highlights the importance of improving treatment and providing ongoing support for individuals with ADHD who struggle to control attention and behavior, as it suggests that the long-term impact on outcomes can be meaningful and substantial.

Despite a wealth of research and attention, ADHD remains in many ways a misunderstood condition, especially for women and girls. Researchers, clinicians, and the broader public have shifted to some degree on the notion of ADHD as something that debilitates to something that differentiates. Sabrina Park, writing in 2019, chronicled 11 women with ADHD whom she described as "thriving:" including actress Emma Watson who also graduated from Brown University and served as a United Nations Goodwill Ambassador, vocalist Solange Knowles who lamented people's perceptions of her before her diagnosis, Olympic gold medalist Simone Biles, and even a YouTuber who created a "How to ADHD" informational channel (Park, 2019). Entrepreneur magazine similarly chronicled ADHD success stories, such as JetBlue Airways founder Devid Neeleman and Virgin Group founder Richard Branson (Belanger, 2017). That article references research in which Wiklund et al. (2016) studied 14 entrepreneurs previously diagnosed with ADHD, noting that impulsivity can be a major driver of entrepreneurial action, and hyper-focus can spur such action to consequences. Regardless of these specific cases or potential benefits to ADHD, we acknowledge that much of the western world is incompatible with the cognitive and behavioral profile of ADHD. Thus, in this work, we explore technologies that seek to support people with ADHD in learning to accommodate the environment around them as well as technologies that seek to change the environment to accommodate them. As Harvey Blume (1998) noted, "Neurodiversity may be every bit as crucial for the human race as biodiversity is for life in general. Who can say what form of wiring will prove best at any given moment?"

1.2   THE EVOLUTION OF "TECHNOLOGY" IN MENTAL HEALTH

Although the vast majority of technology profiled in this book is more recent, it was worth understanding to some degree the origins of the notion of the use of technology in mental health. This has, in many ways, tracked the use of technology in other parts of healthcare, but the trends are distinct in interesting ways, with a much stronger emphasis on self-instruction and home care than physical health trends.

In the 1970s, as hospitals and other health systems computerized, mental health also began to include computerization. For example, Johnson and Williams (1975) described an "on-line computer technology" for mental health admissions. This simple system collected psychological, social, and physical information that could then be used for clinical decision-making. Other systems would follow, and some have even been adopted and put into widespread clinical use. And yet, many of our systems still rely on paper and pencil, particularly for smaller practices. A 1976 review of the state of the art in computers and mental health care delivery (Johnson, Giannetti, and Williams, 1976) predominantly described large-scale data collection and clinical decision support systems.

As in other areas of healthcare, it would take the advent of personal computing to push the notion of technology in mental health into more mainstream practice. An instructional book, complete with included CD-ROM from the late 1990s, describes "practical advice" for even the most "technophobic" clinicians and researchers (Rosen and Weil, 1997). Yet, even then, much of the literature from this time period focused on screening and assessment (e.g., Munizza et al., 2000; Puskar et al., 1996; Stein and Milne, 1999) and computerized decision-support (e.g., Knight, 1995) as opposed to using technology directly for treatment, which was still an emerging area of research (e.g., Budman, 2000; Huang and Alessi, 1998; Riemer-Reiss, 2000). A focus, then as now, was on the notion of treating patients and providing support services at a distance.

In more recent years, the notion of technology for mental health, and for ADHD in particular, has gained enough mainstream interest that a myriad of smartphone "apps;" games for computers, phones, and even in Virtual Reality; and other approaches have become commonplace. Some of this is driven by innovative clinicians, by people with ADHD themselves, and by friends and family members of people with ADHD, making the effects disparate and hard to measure. Several reviews of these technologies and their evidence exist (e.g., Batra et al., 2017; Grist, Porter, and Stallard, 2017; Hansen, Broomfield, and Yap, 2019; Hollis et al., 2017; Michel, Slovak, and Fitzpatrick, 2019; Torous et al., 2018). These explorations can point us toward a variety of future directions with creative solutions and point to the need for more empirical research (e.g., Comer and Myers, 2016; Firth et al., 2018; Ralston, Andrews, and Hope, 2019). Such research also highlights risks and obstacles to technological approaches (e.g., Bhuyan et al., 2017) but cannot yet, in most cases, provide solid extensible guidance and recommendations.

Technology and ADHD have a complicated relationship. While researchers, clinicians, and educators have sought to use technologically enabled approaches to support people with ADHD, others have sought to detect a potential connection between the rise in use of computerized technologies and ADHD itself (e.g., Visser et al., 2014; Beyens et al., 2018). Despite the widespread problematizing by the scientific community of claims that screen and media use cause a wide range of psychiatric disorders (Odgers and Jensen, 2020; Stiglic and Viner, 2019), these myths continue to be perpetuated by both the popular press and an increasingly limited number of researchers (e.g., Twenge et al., 2018). These claims are problematic in part due to their limited statistical power but largely due to the implicit ableism and offensive nature of claims that assert technology has "destroyed a generation" (Twenge, 2017) largely related to claims that this new generation has and makes different choices than was true of prior generations, particularly in relation to independence and social engagement. By asserting that ADHD is caused by screen time and media use (e.g., Tamana et al., 2019, Rosenblatt, 2019) and that, therefore, all children should have such screens removed from their lives or have greatly reduced access (Twenge and Campbell, 2018), we implicitly assert that living with ADHD is worse than living without educational opportunities, connectivity to remote social opportunities, and connection to the larger culture. We are writing this book at a particular moment in time, during COVID-19 stay-at-home orders and subsequent social distancing, and this experience reinforces the deeply problematic nature of staking these claims that are not backed in science and that contribute to bias and problematic policies.

Further challenging reviews and thoughts about Mental Health Technology, and specifically those for use surrounding ADHD, is the use of the word "technology" itself by the interdisciplinary mix of people working in this space. Some clinicians and mental health researchers use the word technology to mean concepts, such as when Swift and Levin defined "empowerment" as an emerging mental health technology (Swift and Levin, 1987). In this case, they were referring to the feeling of power and competence as well as the modification of the structural conditions to invoke these feelings. In this case, and others like it, the authors appear to be using the term technology to describe any tool. Similarly, Byrnes, and Johnson (1981) use the term technology to describe a new process. At the far other end of the spectrum, for many computer scientists and engineers, technology is an evolving concept, and the term itself is difficult to pin down at times, describing only digital or electrical solutions and at others more traditional tools, such as paper as technology. This concept grows murkier when one considers the publication and funding realities of research in the information and computer sciences, in which only new radically cutting-edge solutions are considered worthy of scientific exploration. In this case, the word technology may often be preceded by "novel" or "innovative."

1.3   OTHER RELEVANT ADHD AND TECHNOLOGY REVIEWS

This book is not the first effort to catalog research related to ADHD and technology in some form or another. However, prior reviews published in scientific journals have primarily focused a specific form of technology and its impact on targeted outcomes in ADHD. For example, Powel et al. (2017) focused on the ten most popular smartphone applications targeting children and young people with ADHD or their caregivers and noted the lack of an evidence base for the applications, as well as concerns about their utility and inadequate privacy/security features. They observed that once an application had been purchased and downloaded, there was little opportunity for feedback on utility; participants interviewed in their research identified a range of limitations in the available products, suggesting that they had fallen short when it came to addressing patient, family, and clinician needs. Powel et al. (2018) conducted a second review of 14 studies focused on technologies designed to encourage and support self-management for individuals with ADHD. They noted that although the research indicated some promise for technology to support self-management of ADHD-related difficulties, research was limited by small sample sizes, weak outcome measures, and limited integration of psychoeducation. As another example, in a recent preprint, Hussain

(2020) catalogs the design methods, sample size, and specific subdomains of 27 papers from the ACM Digital Library between 2007 and 2017 focused on technology for ADHD.

As we describe in Chapter 4, for example, hundreds of journal articles and several systematic and meta-analytic reviews have been published addressing computerized cognitive training (e.g., attention training using a computer program designed to look like a video game) for individuals with ADHD. These works have contributed to a growing understanding of how technology may or may not ameliorate ADHD symptoms and improve outcomes. Our intent in this book is to take a broader perspective and describe a wider range of technologies that have been addressed in prior individual reviews. When relevant to a particular chapter, we cite those other review articles and hope that readers may engage with those works as they find them useful and relevant to their own work.

This book also is not meant to include every article ever written about ADHD in relation to technology or technology in relation to ADHD. For one, the book would never be complete as research is ongoing and growing in this space. More importantly, however, scientific journals are the appropriate space for systematic reviews or structured meta-reviews and meta-analyses, such as those described in the previous paragraph. This book is meant, instead, to lay out an intellectual space. The hope of this approach, as has been in true in many volumes in this series, is three-fold. First, we seek to educate and support scholars choosing to move into this area. Second, this book should inspire others to build on, critique, and update the exciting advances we overview here. Finally, this book can serve as a resource for those in the industry looking to develop a new product or modify an existing one to provide a better user experience for someone with ADHD, to enable people with ADHD to make their own content and products or to develop new assessment, diagnostic, and therapeutic tools.

1.4   STRUCTURE OF THIS BOOK

This book is organized by functional domain within the broader context of living with ADHD. While we do not focus on any particular age group, that organization means that some chapters are more likely to focus on children while others may be more centered around the adult experience. We begin in the next chapter with a description of our initial structured literature review and the themes used to classify the literature in that review. This classification scheme runs throughout the book and provides the backbone for the remaining chapters. Chapter 3 focuses on the diagnosis and assessment of ADHD, a series of tools and approaches that cross many of these functional domains. Chapter 4 focuses on cognition and attention and includes a discussion of tools developed to train these skills in individuals with ADHD. In Chapter 5, we focus on social and emotional skills, including the experiences of both children and adults. In Chapter 6, we review tools to support behavior management and self-regulation. Chapter 7 is dedicated to academic skills and so is heavily tied to the experience of childhood schooling as well as higher education. By contrast, the next chapter, Chapter 8, highlights the role of technology in supporting everyday life skills and employment and is thus more focused on the experiences of young adults and adults. Chapter 9 describes technologies to support and improve motor skills, physical access, and physical behaviors. Finally, we conclude with a discussion of where we are headed as a field and what opportunities may lie ahead. Of course, many projects overlap in categories and so can be found in multiple chapters. Likewise, points from the discussion are relevant to detailed portions of the individual chapters as well. Thus, readers may wish to begin with the discussion, read detailed chapters of interest, and return to the final chapter again.

Within each chapter, we describe the background of the experience of ADHD as related to the overall topic of that chapter. In these cases, we highlight both strengths and challenges, indicating the ways in which technology can help people with ADHD to improve their life experiences, ways that technology might help people without ADHD to be supportive and inclusive, and ways in which the currently available technological solutions do not appear to be effective. Finally, we close each chapter with some indication of the conclusions we can draw from inspecting this literature together and the future directions the field might wish to pursue.

CHAPTER 2

Methods and Classification Scheme

In this chapter, we provide a description of the methods we used to identify and analyze interactive technology research included in this book. Based on the existing work of Kientz et al. (2019) in classifying papers in the broad space of interactive technologies for autism, we modified their classification scheme to help categorize each technology approach. This approach is particularly appealing in that it is both descriptive and explanatory. Using this framework, we have been able to trace the history and evolution of this research within specific technology verticals. This approach then provides a roadmap for future research by showing both where the field is headed and where it perhaps could or should be with some appropriate intervention by researchers, practitioners, product developers, or even funding bodies.

2.1 METHODS

Given the rapid growth rate in this area of research, the ambiguous definition of technology, and the rapid changes in clinical and educational practice, as in the Kientz et al. volume that was focused on autism, we do not represent this work as a complete review of the literature but rather an overview of the preponderance of evidence for and open questions about various technological approaches. Notably, there are numerous off-the-shelf, open-source, and commercial products that serve similar needs, either explicitly or through appropriation and adaptation. We intentionally exclude applications from popular media, such as games for children with ADHD found in the Apple App Store or on Google Play. These marketplaces are rich with different applications, but in general, they are beyond the scope of this book unless they have been studied in the scientific literature. Given our focus on research, this book is limited to research projects and research-validated products with some mention of other non-research products when particularly relevant.

We conducted our research for this book in two phases. In the first phase, we conducted a structured literature review using searches in PubMed, ACM Digital Library, and IEEE Xplore for articles published in English from 2004 to June 1, 2020. We witnessed an inflection point in the ACM and IEEE articles, in particular around 2004 (see Figure 2.1). Search terms included: "ADHD," "Attention Deficit Hyperactivity Disorder," "treatment," "digital," "intervention," "assistive technology," "computer intervention," "computer assisted," "sensor," "mobile," "wearable," "smartphone," "tablet," "robot," "virtual reality," "augmented reality," "neurofeedback," "working memory training," "cognitive training," "internet," and "web." We limited results to published peer-reviewed research papers, excluding research published solely as abstracts or extended abstracts; these research abstracts were excluded during the first phase of screening (Figure 2.2).

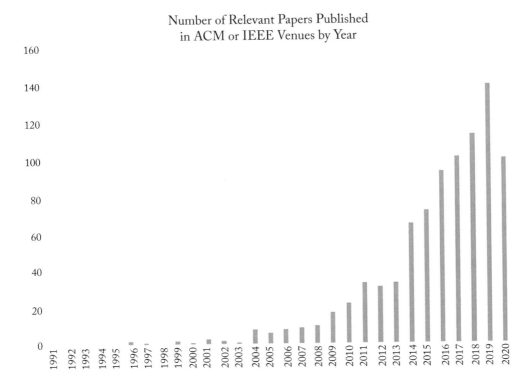

Figure 2.1: Chart graphing publications in the ACM and IEEE digital libraries using "ADHD" as the keyword starting in 1991 shows 2004 as a clear point after which the topic began to gain substantial interest to technology and computing researchers. Notably, at time of publication, 2020 is still in progress, hence limiting the numbers of publications available for that year.

We included research articles that used consumer, clinical, and educational technologies. To identify this initial set of literature for the book, we used the PRISMA process for identifying appropriate articles for inclusion (see Figure 2.2, which demonstrates how we identified papers focused on the application of technology to intervention and treatment only).

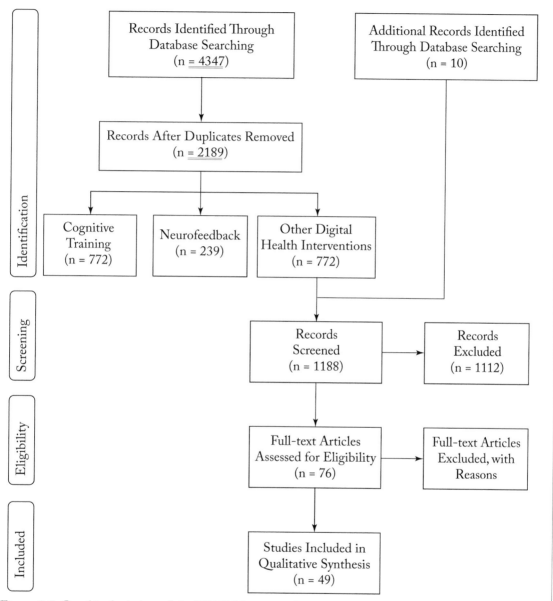

Figure 2.2: Graphic depiction of the PRISMA process for initial meta-review.

Once this initial corpus was developed, we broadened our search to include additional conferences and journals and additional topics in an iterative fashion. We checked the reference lists in the papers in our initial corpus and used Google Scholar to scan the papers that cited them. We then analyzed these papers using the classification scheme outlined below. Finally, when we identified holes in the literature through this analysis, we searched again, this time using Google Scholar with the broad sets of terms outlined above but expanding to other databases, such as JSTOR and EPIC that are indexed by Google. This process continued iteratively as we developed the text for the book itself. Thus, while we do not compile a comprehensive view across all potential fields that intersect ADHD and Technology, we are confident in our ability to detail the landscape of studies to provide a foundation for scholars interested in expanding research in this area as well as practitioners or developers looking to build on what already has been done.

2.2   CLASSIFICATION SCHEME

We adapted the framework developed by Kientz et al. (2019) to organize the existing literature. This adapted framework consists of five dimensions along which projects can be categorized. Within each dimension, we determined several labels within that dimension that could describe the work, again derived in part from the Kientz et al. framework. Below we describe the five dimensions, along with the associated labels within those dimensions and their operational definitions. In many of these dimensions, it is possible that multiple sub-labels apply. For the purposes of the readability of this book, we categorize them roughly in the most dominant sub-label, noting the overlap, however in Table 2.1.

2.2.1   INTERACTIVE TECHNOLOGY PLATFORM

This category describes the primary platforms, form factors, or delivery mechanisms used by the technology or application.

- **Personal Computers and Web:** Includes applications that use a traditional keyboard, mouse, and monitor, and Internet-based applications that are primarily designed for access via a desktop-based web browser. Can also include laptop-based technologies, but the primary differentiator is those that are intended to be stationary and not mobile.

- **Mobile Devices and Tablets:** Includes applications delivered on cell phones, PDAs, tablets, or other mobile devices intended for personal use. Can be used in multiple environments or anywhere the user goes.

- **Sensor-Based and Wearable:** Includes the use of sensors (e.g., accelerometers, heart rate, microphones, brain-computer interfaces (BCI) etc.), both in the environment and on the body, or computer vision to collect data or provide input.

- **Virtual and Augmented Reality:** Includes the use of virtual reality, augmented reality, virtual worlds, and use of virtual avatars.

- **Robotics:** Includes physical instantiations of digital interactions. Includes both humanoid or anthropomorphic robots and general digital devices that carry out physical tasks. Includes both autonomous robots and those operated remotely by humans.

- **Natural User Interfaces:** Includes the use of input devices beyond traditional mice and keyboards, such as pens, gestures, speech, eye tracking, multi-touch interaction, etc. Interacts with a system rather than just providing passive input. Although this is a term of art for computing researchers, it is not without its critics, as the term "natural" here largely means "not mouse and keyboard" rather than anything that is truly "natural" in the sense of not being artificial, synthetic, or learned.

2.2.2   DOMAIN

This category refers to the area of functioning within individuals with ADHD that the technology targets, such as helping with the acquisition of certain skills or addressing particular deficits.

- **Cognition and Attention:** Includes applications designed to improve cognitive skills, including attention, working memory, and other executive functions.

- **Social and Emotional Skills:** Includes technologies designed to improve social skills and peer and family relationships and reduce emotional symptoms, such as anxiety and depression.

- **Behavior Management and Self-Regulation:** Includes applications focused on supporting self-regulation and behavioral approaches, such as technologies to track or deliver token economies.

- **Academic and Organizational Skills:** Includes applications focused on time management, organization, reading, mathematics, and other academic skills.

- **Life and Vocational Skills:** Includes applications or technologies designed to support the development of independent living skills and other specific skills, such as training to support safe driving behaviors.

- **Motor Behaviors:** Includes applications designed address hyperactivity and motor coordination skills, such as tools to help improve fine motor coordination.

2.2.3   GOAL

This category refers to the primary goal of the technology itself. Some technologies related to ADHD are intended to support a diagnostic process or to screen for symptoms of ADHD, whereas others are intended to support or deliver interventions.

- **Treatment/Intervention:** Includes (1) applications that attempt to improve or produce a specific outcome in an individual with ADHD. May focus on teaching new skills, maintaining or practicing new skills, or changing behaviors; and (2) applications that provide support for caregivers, educators, clinicians, and other professionals to further their understanding of ADHD and intervention strategies or improve their skills as caregivers.

- **Diagnosis/Assessment:** Includes (1) applications that help either screen for ADHD diagnosis in the general population or test for symptoms of ADHD in clinical settings; (2) applications focused on the collection and review of data over time to assess an individual's learning, capability, or level of functioning. The data collected is intended for end users and/or people caring directly for individuals with ADHD; and (3) applications or projects that use technology in the collection and analysis of data by researchers to understand more about ADHD and its features or characteristics. Tools in this category generally are not yet available for or may not be appropriate or feasible for home or community use.

2.2.4   TARGET END USER

This category focuses on the person or persons who will actually be interacting with the technology itself and are considered the primary users. It does not refer to secondary stakeholders or those who may benefit from the technology but do not actually interact with or use it.

- **Child with ADHD**

- **Adolescent with ADHD**

- **Adult with ADHD**

- **Family Caregiver:** Includes anyone who is not a professional who cares for or supports an individual with ADHD. May include parents, siblings, other family, friends, volunteers, etc.

- **Provider/Researcher:** Includes a paid professional who works with individuals with ADHD. May include teachers, medical professionals, doctors, occupational therapists, physical therapists, speech therapists, applied behavior therapists, or other allied health professionals. Also includes a person intending to collect data or conduct studies about individuals with ADHD and publish something generalizable about the data. This does not include researchers running a study about technology for an individual, only when the researcher themselves is one of the technology's primary users.

2.2.5   SETTING

The care of individuals with ADHD takes place in a number of different settings. This category refers to the settings or locations in which the technology is primarily intended to be used.

- **Home:** The home or personal living space of a person with ADHD and/or their family.

- **School:** A public or private place for educating individuals with ADHD. Includes both schools that specialize in ADHD education as well as general, inclusive classrooms. Could include all levels from pre-school through postsecondary education.

- **Research Lab:** Technology intended for use in a research laboratory under careful observation or that has been tested only in controlled settings.

- **Clinic:** A place of professional practice that is not intended for education, such as a doctor's office, therapist's office, or a specialty service provider.

- **Everyday Life:** Including work, transportation, socialization, shopping, and other daily activities.

2.3   APPLYING THE CLASSIFICATION SCHEME

To illustrate how we applied this framework, here we describe a single paper and how it was classified. Weisman et al. (2018) published a study evaluating the ICON smartphone application and illustrated how innovative technologies could be applied to improve medication adherence, which can be a substantial problem in ADHD treatment. We classified this work as a mobile device (technological platform) that targets organizational skills (domain) in order to support intervention and patient education (goals). The application targeted children and their caregivers (target end users) and was developed for use at home (setting). Also, Table 2.1 includes 68 projects on technologies reviewed for this book and their associated coding within the classification scheme. We chose the 68 papers based on their diversity across specific areas to exemplify the use of the classification scheme.

In the subsequent chapters, we use this framework to describe the approaches to interactive technology research in the ADHD space. Although we adapted the Kientz et al. (2019) framework for this analysis, we structure the rest of the book differently. Where they structured their review based on the technological platform as the primary organizing principle, we instead focus on the domain of concern that the design, technology, and/or research project was developed to address. Subsequent chapters include Diagnosis and Assessment of ADHD Symptoms (Chapter 3), Cognition and Attention (Chapter 4), Social and Emotional Skills (Chapter 5), Behavior Management and Self-Regulation (Chapter 6), Academic and Organizational Skills and Support (Chapter 7), Life and Vocational Skills and Support (Chapter 8), and Motor Behaviors and Physical Accessibility (Chapter 9). Some applications and technologies might fit into more than one of these categories, in which case they are mentioned in both relevant chapters and cross-referenced appropriately.

Within each chapter, we describe the challenges and opportunities of the specific domain as well as the interventions and innovative technologies that have been developed in this space. We describe solutions that have been developed or used by people with ADHD and the ways in which these technologies themselves can be challenging or problematic. Finally, we close with potential future opportunities in this area of research.

Any book will necessarily only provide a review that is a snapshot in time. Both our understanding of ADHD and the technologies we use to support our lives are ever-changing. We hope this book provides a foundation upon which others may build. Additionally, we expect that others can use the methods we outlined above to monitor additional literature as it is published and stay up to date. Finally, we hope that people with ADHD themselves, whether researchers or not, will take up this work, apply to it their lives, and inform our future work with their lived experiences.

Table 2.1: **Coding of 70 projects used to test our classification scheme**

	Technology Platform						Domain						Goal		Target End User					Setting				
	Personal Computers and Web	Mobile Devices or Tablets	Sensors/Wearables/EEG	Virtual and Augmented Reality	Robotics	Natural User Interface	Cognition/Attention	Social/Emotional Skills	Behavior Management/Self-Regulation	Academic/Organizational Skills	Life/Vocational Skills	Motor Behaviors/Physical Activity	Treatment/Intervention	Assessment/Diagnosis	Child with ADHD (<12 years)	Adolescent with ADHD (13–19 years)	Adult with ADHD (>20 years)	Caregivers (Parents/Teacher)	Providers/Researchers	Home	School	Research Lab	Clinic	Everyday Life
Aase and Sagvolden, 2005, 2006	*								*					*	*							*		
Adams et al., 2009				*			*							*	*	*						*		
Alchalabi et al., 2018; Alchalcabi et al., 2017			*				*							*			*					*		
Areces et al., 2018a, 2018b, 2019				*			*							*	*	*						*		
Asiry et al., 2018	*						*			*				*	*				*					
Avila-Pesantez et al., 2018	*			*	*		*						*		*							*		
Babinski and Welkie, 2019		*						*					*			*		*		*				
Bashiri et al., 2018	*						*						*					*					*	
Benzing et al., 2018; Benzing and Schmidt, 2017, 2019					*	*	*					*	*		*					*				
Breider et al., 2019	*									*			*					*	*	*		*		
Bruce et al., 2017	*										*		*			*	*							*
Bul et al., 2015, 2016, 2018	*						*	*		*			*		*					*				
Chen et al., 2018		*	*			*	*		*					*	*							*		
Cibrian et al., 2020a		*	*						*				*		*	*				*	*			
Clancey, Rucklidge, and Owen (2006)				*							*		*		*									*
Clark-Turner and Begum, 2017					*				*	*					*	*							*	
Corkum et al., 2019	*						*		*				*					*				*		

Davis, et al., 2018	*					*					*	*				*					
Díaz-Orueta et al., 2014; Iriarte et al., 2012			*			*		*							*	*	*				*
Dovis et al., 2015; Prins et al., 2013	*							*			*	*				*					
DuPaul et al., 2018	*							*			*				*	*					
Eliasson et al., 2004		*								*	*	*									
Eom et al., 2019			*			*					*	*	*						*		
Faedda et al., 2016		*						*			*	*	*			*	*				
Fang et al., 2019			*			*	*	*			*	*							*		
Farran et al., 2019			*								*	*	*	*	*	*	*				
Fujiwara et al., 2017		*	*					*			*			*	*						*
García-Baos et al., 2019					*	*					*	*	*			*					
Garcia-Zapirain et al., 2017					*	*					*					*			*		
Hyun et al., 2018		*				*					*	*				*					
Jaiswal et al., 2017	*				*						*	*			*	*					
Jimenez et al., 2016				*				*			*	*						*			
Johnson et al., 2010		*	*								*	*	*			*					
Kam et al., 2010			*								*	*	*			*					
Kanellos et al., 2019		*	*		*	*					*					*				*	
Kim et al., 2020			*			*					*		*			*					
Kollins et al., 2020		*				*					*		*			*					
Lewis et al., 2010	*									*	*		*			*					
Luna et al., 2018		*		*						*	*		*					*			
Matic et al., 2014	*	*				•				*		*		*				*			
Mautone et al., 2005	*									*	*		*			*					
Mock et al., 2018		*				*					*	*	*			*					
Muñoz-Organero et al., 2018, 2019			*								*	*	*			*					
Neguţ et al., 2017			*			*					*	*	*		*			*			
Nolin et al., 2016			*			*					*	*	*		*			*			
Olthuis et al., 2018	*	*						*			*				*	*					
Palsbo and Hood-Szivek, 2012					*						*	*		*				*			
Park et al., 2009	*		*			*	*				*	*				*	*				
Parson et al., 2007				*		*					*	*						*			
Pina et al., 2014		*						*			*				*	*					
Pollak et al., 2009			*			*					*	*	*								
Rijo et al., 2015	*					*					*		*								
Rodríguez et al., 2018			*			*	*				*	*	*				*				
Ryan et al., 2015	*							*			*				*	*					
Santos et al., 2011	*					*					*	*	*			*					*

	1	2	3	4	5	6	7	8	9	10	11	12	13	14	15	16	17	18	19	20	21
Schuck et al., 2016		*							*			*		*			*			*	
Sehlin et al., 2018; Söderqvist et al., 2017; Wentz et al., 2012	*	*						*		*	*	*		*	*	*	*	*			*
Shema-Shiratzky et al., 2019			*			*	*				*	*		*						*	
Shih, 2011; Shih et al., 2014, 2011	*		*								*		*	*				*			
Sonne et al 2016a, 2016c		*							*	*		*		*		*		*			
Sonne et al., 2015	*		*				*					*		*					*		
Sonne and Jensen, 2016; Sonne et al., 2017			*						*			*		*						*	
Spitale et al., 2019		*								*			*		*					*	
Tam et al., 2017					*						*		*	*							*
Tan et al., 2019			*	*			*						*	*						*	
Thomas et al., 2010		*							*	*			*	*		*		*	*		*
Toshniwal et al., 2015		*								*			*		*		*		*		
Wills and Mason, 2014		*					*		*		*			*					*		
Wood et al., 2009			*								*		*	*	*					*	
Yeh et al., 2012			*				*						*	*						*	

CHAPTER 3

Computationally Supported Diagnosis and Assessment

ADHD is highly individualized, with each child, adolescent, and adult expressing symptoms in different ways, with no clear physical test to diagnose the condition. Rather, diagnosis is typically given in response to a series of behavioral observations and reports sometimes combined with *neuropsychological* assessments and self-report of internal feelings. Many children and adults with ADHD go years or even lifetimes without a formal diagnosis. At the same time, there is growing interest in increasing the rigor of diagnostic procedures as well as the assessment of progress in response to a variety of interventions. These kinds of tests, requiring intensive support from clinical and educational resources, can be difficult to scale, but with the prevalence of ADHD such scale is necessary. Thus, technology offers an opportunity to support human professionals and experts in their diagnostic and assessment work. In this chapter, we first introduce the basics of ADHD diagnoses, followed by the extensive literature on computational approaches to augmenting diagnostic and assessment work.

3.1 ADHD DIAGNOSIS

No single diagnostic system or procedure for ADHD has been discovered or adopted yet, and a rigorous evaluation process often involves hours of clinician time. Thus, technological tools have been developed to glean diagnostic information useful to clinicians evaluating patients for ADHD. On the one hand, research has often focused on classifying behaviors or physiological data between individuals with ADHD and neurotypical peers. On the other hand, some research has focused on finding patterns among the data to generate symptom profiles of ADHD.

For a number of years, computers have been used to test attention, yielding normative data that can be useful as part of a comprehensive diagnostic evaluation for ADHD. One of the most commonly used technological tools in clinics is the Conner's Continuous Performance Test, currently in its 3rd edition (www.mhs.com/CPT3). This task-oriented computerized test of attention allows the evaluation of an individual's ability to maintain attention across a period of time and produces standardized, normative scores that can help identify difficulties in attention that are sensitive to development. Similarly, the Test of Variables of Attention (TOVA: www.tovatest.com) also tests attention and produces informative scores. Both tests assess attention using an intentionally boring "computer game" designed to evaluate how well an individual can regulate attention over time. Neither claims to replace a comprehensive diagnosis, but both are widely thought to generate useful information to inform a comprehensive diagnostic assessment.

As digital recording technologies become more pervasive, more robust, and less costly, it should come as no surprise that they are increasingly used to support medical diagnoses of both mental and physical health conditions. Mental health, in particular, has been traditionally difficult to diagnose because few definitive tests exist for certain behavioral health conditions. Recent research projects indicate that diagnoses of ADHD may be greatly enabled and streamlined through automated sensing and monitoring of physical health and observable behaviors, in particular, as they relate to potential co-occurring motor conditions. Additionally, the use of these tools has, in some cases, enabled researchers to rule out such connections from conditions that only appear to co-occur in clinical practice but such connections were not able to be tested rigorously previously. In this section, we describe the extant research in using technology to understand and support the diagnosis of ADHD, including studies that found use of such technologies to be promising as well as those that demonstrated certain behaviors or symptoms either cannot be effectively measured or are not predictive of diagnoses as expected.

3.2   COMPUTATIONAL DIAGNOSTIC AND ASSESSMENT APPROACHES

Three primary approaches have been explored in research attempting to create non-biased diagnostic tests for ADHD. Computational techniques can be used to classify data from brain activity, either EEG or fMRI. Data can also be collected from sensors (on the body, in the environment, or inherent to computational tool use) used during everyday activities and then create computational models that can classify unseen data instances. Finally, some approaches have been focused on designing and developing serious games or environments where users can play and interact. The interactions of the users with the game are analyzed to infer if the user has ADHD or related symptoms

3.2.1   COMPUTER AND MOBILE PHONE-BASED ASSESSMENTS

Given that ADHD is largely diagnosed by clinician observation and report, parental report, self-report, and behavioral diagnostics, it should be no surprise that one role computation has taken in this process is simply to document, support, and analyze such collected data. Clinical decision support systems (CDSS) (Berner, 2007) can improve healthcare outcomes by facilitating evidence-based medical practice. For example, Kemppinen et al. (2013) studied the implementation of a customized CDSS for adult diagnosis of ADHD, finding that implementation of such a system required the establishment of a standard for the care of adult ADHD patients, enabling quick changes as new information is incorporated into their processes.

Some computerized tests support the patients directly, rather than as part of a clinical decision support structure. For example, MATH-CPT uses an on-screen sequence of simple mathematical questions, which the authors hope can be used to diagnose participants with ADHD. In a study of 303 participants (63 with ADHD), MATH-CPT correctly classified 91.6% of participants, performing better in their study than the Test of Variables of Attention (TOVA) (Rodríguez et al., 2018). QbTest, a computerized continuous performance test, has been shown to reliably differentiate between children with ADHD and those without (Emser et al., 2018; Hult et al, 2018) as well as adults (Bijlenga et al., 2019; Edebol, Helldin, and Norlander, 2013; Emser et al., 2018; Hirsch and Christiansen, 2017; Lis et al., 2010). Similarly, QbCheck was built as an online assessment tool for families who were not able to come into a clinic but wanted some initial screening or assessment for ADHD. In a study with 142 adolescents and adults (69 with ADHD), high convergent validity was observed between QbCheck at home and QbTest (Ulberstad, 2016) in the clinic (Ulberstad et al., 2020). Despite the promise of these results for a notion of automated diagnosis, many researchers have noted that there are just too many limitations inherent to claiming to diagnose a clinical condition without a qualified clinician. Thus, many experts advocate for such tools to provide useful information to the diagnostic process rather than replace it.

Finally, parents are sometimes the target of data collection as part of family and child assessments. For example, Hsieh, Yen, and Chou (2019) examined the relationship between the Parental Smartphone Use Management Scale (PSUMS) and ADHD symptoms, finding that PSUMS accounted well for reactive management, proactive management, and monitoring in relation to parental feelings of self-efficacy. Although PSUMS itself is not delivered via technology, it is a potentially useful tool for clinical practice and parenting groups to understand parental management and point to particular challenge areas around technology use for families. On the other hand, Li and Lansford (2018) used smartphones in the form of ecological momentary assessment (Shiffman, Stone, and Hufford, 2008) to track parents of 184 kindergartners, with and without ADHD, for one week, finding limited relationships between the stress of parenting and variability in harsh parenting behaviors. Inclusion of assessment tools like this into other therapeutic or educational applications for parents may be particularly valuable.

3.2.2   VIRTUAL AND AUGMENTED REALITY TO ASSESS ATTENTION

Virtual reality (VR) provides an option for creating dynamic, immersive three-dimensional environments in which behaviors can be recorded and analyzed for later assessment more easily and in more detail than in a traditional physical classroom. VR enables the presentation of cognitive tasks that systematically target attention performance, which while only in VR may provide insights into attention performance outside of this environment. Although there are several studies in virtual and augmented reality to assess ADHD (e.g., Fang et al., 2019; Keshav et al., 2019; Knouse et al., 2005;

Silva and Frère, 2011), two main projects have been widely explored, the Virtual Reality Classroom (then called ClinicalVR) and AULA.

The Virtual Reality Classroom has been used to assess the attention of children with ADHD for the last two decades (Rizzo and Buckwalter, 1997; Rizzo et al., 2000, 2003, 2004, 2006). The VR Classroom includes a standard rectangular classroom rendering containing three rows of desks, a teacher's desk at the front, a blackboard across the front wall, and a virtual teacher who presents as a woman. On the left side, a large window looks out onto a playground with buildings, vehicles, and people. As a tutorial, the virtual teacher instructs the participants to spend a minute looking around the room and point, name observed objects, and then play a quick game. Following the tutorial, participants can play two games, one focused on identifying letters on the blackboard, with or without audio, visual, and 3D audio/visual distracters, and assessment of the virtual teacher's accuracy at identifying images drawn on the blackboard. The project was revised by digital mediaworks, inc. (http://www.dmw.ca/) and called ClinicalVR: Classroom-CPT (Nolin et al., 2016). The system was then adapted to a Unity 3D version (Yeh et al., 2012; Tan et al., 2019).

Several studies have been conducted to test the efficacy and efficiency of the VR Classroom. VR classroom measurements are consistently correlated with traditional measurements (Parson et al., 2007; Pollak et al., 2009), making it a promising place for testing new interventions. However, children with ADHD were more affected by distractions in the VR classroom than those without ADHD (Adams et al., 2009). In the newest version, ClinicalVR, studies show that assessments made by monitoring behaviors are reliable and unaffected by gender, while age is impacted (Nolin et al., 2016). Although there are differences between performing these tests in the virtual environment from a more traditional computerized one, measurements were able to distinguish between children with ADHD and neurotypical children (Negut et al., 2017), provide incremental validity beyond that of teacher and parent report of behavior (Coleman et al., 2019), and add social cues to the assessment (Eom et al., 2019).

Similarly, The AULA Nesplora or just AULA has also been used to assess ADHD in children (Climent and Banterla, 2010). AULA is a virtual school classroom where children used a head-mounted display with movement sensors, earphones, and a single-button switch to interact in the environment. AULA had two main activities: a NO-X paradigm-based exercise (i.e., "Press the button when you DO NOT perceive the target stimulus") and an X paradigm-based exercise (i.e., "Press the button whenever you DO perceive the target stimulus"), that children can play for 20 minutes (Díaz-Orueta et al., 2014). This approach has also been explored using an aquarium environment (Camacho-Conde and Climent, 2020).

Studies have been conducted with children to first correlate AULA measurements with more common and standard Continuous Performance Tests (CPT), confirming the validity between the tests (Díaz-Orueta et al., 2014). AULA was able to differentiate between children with ADHD and without pharmacological treatment for a wide range of measures related to inattention, impulsivity, processing speed, motor activity, and quality of attention focus (Díaz-Orueta et al., 2014; Iriarte et al., 2016). The cognitive scales used in AULA take into account the virtual environment to better characterize the difficulties encountered by people with ADHD (Areces et al., 2018a), VR is able to classify the impulsive/hyperactive, inattentive behaviors, and combined symptoms (Areces et al., 2018b). AULA predicts current and retrospective ADHD symptoms (Areces et al., 2019), and better differentiates between ADHD and non-ADHD individuals in comparison with the most commonly used CPT, the Test of Variables of Attention (TOVA) (Rodríguez et al., 2018). These results show promise for improving the performance monitoring data collection that many clinicians use to speed their screening, assessment, and even diagnostic work.

Augmented reality has also been explored to assess ADHD symptoms. For example, Empowered Brain is an augmented reality communication aid for children with ASD. In one study of Empowered Brain, seven high school students diagnosed with ASD played Empowered Brain for one week. In this study, Empowered Brain in-game performance correlated with ADHD symptom severity in students with ASD. This was a relatively small sample but showed promise in using tools designed to support children with ASD in simultaneously assessing them for ADHD (Keshav et al., 2019).

3.2.3   COMPUTATIONALLY ASSESSING MOTOR SKILLS

Using systems that children already engage in within their daily lives that increasingly include touchscreens and other computational tools, makes building assessment into the background appealing. For example, Mock et al. (2018) used touch trajectories during a multiple-choice format math test to predict ADHD. For regression of overall ADHD scores, their mean squared error was 0.096 on a four-point scale ($R^2 = 0.567$). They demonstrated classification accuracy for increased ADHD risk appropriate for use alongside other clinical diagnostics but not on its own (91.1%). Although far from being ready to support ADHD clinical work, as of now it has not been tested with people with ADHD, the Snappy App (Young et al., 2014) and its gaming "in the wild" counterpart (Craven et al., 2014) show some promise in terms of measuring ADHD-related symptoms. These approaches are novel and innovative and should be explored further with both children and adults with ADHD diagnoses or who self-report ADHD symptoms and can be diagnosed as part of a rigorous research study.

VR can be a particularly appealing platform for testing motor, spatial, and physical abilities because it can represent a fully controlled large area in a relatively small physical space. Additionally, data collected by wearable sensors and the VR environment itself provide vast amounts of data with which to run diagnostic and research tests. For example, in a study that included 51 children with ADHD, Farran et al. (2019) used a VR spatial navigation task to test hypotheses around co-occurring motor impairment in people with ADHD. In this work, they found no relationship between the motor and spatial domains and children with ADHD.

Despite the limits of existing work, VR may be particularly appealing as an assessment and diagnostic approach because it can mimic the natural, physical world while still allowing for an immense amount of control. It, in essence, becomes an experimentally controlled condition. VR can, of course, be fun and engaging. It is not without its drawbacks, however. Even though VR has become more accessible, less expensive, less heavy, and more tolerable (e.g., creates less nausea), it is still not particularly intuitive for many people and may be totally out of reach for people with certain kinds of sensory challenges, including small children and those with neurodevelopmental disorders. As VR advances, we would expect these kinds of applications to improve, particularly to support tools for assessment and long-term tracking of symptoms.

3.2.4   DIAGNOSTIC AND ASSESSMENT GAMES

Assessment and diagnosis of ADHD can be a tedious and repetitive process where children may be easily distracted by a lack of motivation or engagement in the tasks. Those behaviors may be confused with distractions and inattentiveness (i.e., executive dysfunction) associated with ADHD. To create a more engaging platform for assessment, serious games have been proposed. The gameplay dynamic of the serious game can be used to assess ADHD symptoms. For example, The Supermarket Game is a labyrinth that must be traversed while the players acquire items shown in his/her shopping list (Andrade et al., 2006). The game used data mining algorithms to classify the data produced by the game (e.g., points, time). Eighty students (39 with ADHD) played the game. A Naïve Bayes algorithm was able to classify adolescents as with ADHD or without ADHD with a sensitivity and specificity of 70% (Santos et al., 2011). Similarly, PANDAS, is a tablet-based game that allows clinicians and researchers to collect data on gameplay, that is then fed to a linear binary support vector machine (SVM) that was able to classify children with ADHD correctly 86.5% of the time (Mwamba, Fourie, and Van den Heever, 2019). These approaches are promising but have not yet rendered more traditional tests obsolete. Entrenched interests in traditional diagnostic approaches as well as a paucity of data backing up these new approaches combine to make it difficult to make such approaches standard in clinical and educational practice.

To address the gaps in data collection merely on gameplay behavior, researchers have also examined serious games controlled with sensors that get data from the user. The data is then analyzed to identify behaviors that may support the screening or assessment of ADHD symptoms. One approach is to create serious games controlled with commercial EEG (e.g., EMOTIV EPOC+, Neurosky). Although commercial EEG usually has fewer electrodes than clinical units, the open-source libraries attached to them make them appealing to develop serious games. The raw signals then can be analyzed using machine learning techniques (similar to the techniques used in Section 3.2.2).

For example, Alchalabi et al. (2017) developed a serious game in which players collect yellow cubes via mental commands of "push" as quickly as possible. The cubes are in a nature-theme environment (e.g., forest) to enhance calm and relaxation. A pilot study with 4 neurotypical players and 4 with ADHD suggested that machine learning models could classify EEG data of individuals with ADHD with up to 96% accuracy (Alchalabi et al., 2018). These approaches require substantially more work to move toward clinical use. First, clinical EEG systems should be opened up through APIs to allow game makers to develop games using their platforms for increased electrical sensing quality. Second, these studies must be conducted at much larger scales to test performance across a wider variety of participants before they have any chance of being accepted as evidence-based clinical tools.

Taken together, this research demonstrates the potential for using gameplay as an assessment tool. In particular, the automated collection of data pertaining to game behaviors and sensors can augment traditional clinical tests and may provide additional insights once enough evidence is collected to support clinical use. The data collected can also be used to develop machine learning models able to classify individuals most likely to be diagnosed with ADHD or to demonstrate specific ADHD symptoms, which can indicate targets for intervention and greatly streamline clinical workload.

3.2.5   VISUAL OBSERVATIONAL METHODS

Computer vision, computers understanding information from processing digital images or videos, offers a promising approach to the diagnosis and monitoring of a wide variety of physical and mental health challenges. Although concerns around undue capture or too much or inappropriate video will likely deter most people from full-time video capture, a variety of less intrusive approaches have been attempted. These are promising because vision sensors do not require the subject of the analysis to do anything different than they might normally nor wear anything particular to capture the data. One example of such an approach uses RGBD (Color + Depth) sensors to gather facial expression analysis and 3D behavior analysis (Jaiswal et al., 2017). Their results indicated classification rates as high as 96% for controls vs. condition in ADHD and Autism groups, which are promising results as an early screener for either autism or ADHD prior to a clinical diagnosis. This work builds on preliminary studies by Hernandez-Vela et al. (2011), who extracted a 3D skeletal model of the human body using RGBD image sequences to find certain body gestures that indicated ADHD in children. Although not as technologically advanced as these approaches, QbTest is robust and commercially available. This system measures hyperactivity, inattention, and impulsivity using head motion tracking alongside a computerized diagnostic instrument (Pelligrini et al., 2020). This approach requires the subject to wear a head tracking headband and does not include facial recognition.

Although not technically a computer vision approach, Lis et al. (2010) demonstrated that infrared motion-tracking could be used to identify higher levels of motor activity in adults with ADHD than those without. This objective measure of motor activity using a novel technological approach was not extended to serve as a diagnostic but helps to understand more about the role of hyperactivity in motor performance, particularly for adults with ADHD.

3.2.6   MULTIMODAL AND PHYSICAL INTERFACES

Mobile technology can allow researchers, educators, and clinicians to assess behaviors "any time" "anywhere," especially for adolescents and adults. In this sense, researchers have used tablets to assess and screen adolescents and adults with ADHD (Hyun et al., 2018; Loskutova et al., 2019).

Hyun et al. (2018) successfully used a digital tablet to conduct the Rey-Osterrieth complex figure test (ROCF), which is a *neuropsychological* assessment tool comprising of drawings that individuals need to copy and draw from memory (Rey, A., 1942) to assess visuospatial perceptual, visio-motor integration, and executive functions (Smith et al., 2007). Thirty adolescents with ADHD and 30 neurotypical controls were asked to copy the drawing onto a Galaxy tablet screen using a wireless pen. Results showed that using a tablet to do the ROCF assessment enabled researchers to uncover different patterns of visuospatial working memory abilities in adolescents with ADHD (Hyun et al., 2018).

Similarly, Loskutova et al. (2019) tested the feasibility of a two-step screening process for adults with ADHD using computerized surveys on a tablet called Talking Survey™. The surveys included a brief symptom checklist and an assessment of Quality of Life. A study with 711 adults (181 screened positive) and 97 physicians showed that clinicians were willing to screen, diagnose, and treat adults with ADHD using the Talking Survey, but they needed additional resources for a full intervention.

A hybrid approach (wearable sensor, intelligent hardware, and mobile application) has also been explored to create a system to support the assessment of the ADHD diagnosis items from the DSM-5. Chen et al. (2018) develop COSA, a contextualized and Objective System to support ADHD diagnosis. COSA assesses ADHD symptoms using three Serious Games: Protect Apple, Catch Butterfly, and Can't Help Eating Candy. Physiological data, movement data, and task-related data allow the assessment of inattention, hyperactivity, and impulsivity symptoms. Clinicians, parents, and children found the use of COSA acceptable in evaluative questionnaires (Chen et al., 2018).

Robotics, which can include a wide variety of human and self-controlled mechanical and digital devices, have also been applied to the study of motor skills in individuals with ADHD. CARBO (CAretakerRoBOt) was developed to address the fine motor skills and touch sensitivity differences that are sometimes observed in children with developmental disorders in order to assist in delivering sensory integration therapy (Chou et al., 2015; Bucci et al., 2014). CARBO is a socially assistive robot with a form factor that encourages users to rub or pet its surface. The convex shell has 67 tactile sensors and LEDs to provide visual feedback to children. The sensors and actuators allow the creation of interactive games at CARBO, such as ColorMe. ColorMe aims for children to get a tactile response from CARBO (Krichmar and Chou, 2018). To win the game, children need to rub the shell in a single direction at a constant speed. During this time, CARBO gives auditory and motion feedback to the children. A deployment study was conducted with 19 children: 5 were diagnosed with ADHD, 13 had diagnoses of ADHD along with other disorders such as depression or anxiety, and 1 was diagnosed with autism. Children played ColorMe using CARBO, and all the tactile interactions were recorded and analyzed. Results showed that children with only ADHD attempted to complete higher levels but had more errors and erratic movements. Children with autism achieved lower levels and performed more slowly and smoothly. Children with anxiety demonstrated more incorrect movements than others. These results indicate that CARBO was sensitive to individual differences in behavior and the investigators anticipated that it had the potential to provide potentially useful diagnostic information (Krichmar and Chou, 2018). The results point to the potential for future research in robotics to help detect and address clinical needs in individuals with ADHD.

Figure 3.1: **CARBO** on the left with children and on the right up close and illuminated. Images courtesy of Jeffrey Krichmar.

3.2.7 WEARABLE AND PHYSIOLOGICAL SENSORS

Wearable sensors, while more intrusive than computer vision or infrared tracking, are increasingly popular as diagnostic tools as their accuracy increases, the price reduces, and both battery life and the comfort of wearing them improves. These tools can also be used to influence diagnostic tools themselves, such as by demonstrating through wearable activity data that overactivity is indeed one of the core features within ADHD combined subtype as clinically understood for years prior (Wood et al., 2009). A wide variety of wearable sensor projects are focused on using a combination of acceleration and velocity sensors to measure, describe, assess, and diagnose ADHD. In this section, we describe those projects as well as those using a broader range of sensors.

Kam et al. (2010) analyzed data from actigraphs, small devices that record activity levels by sensing physical movement, on the non-dominant wrists of 142 school-children (10 with ADHD) for only 3 hours and used decision-tree algorithms to build 2 binary models (ADHD vs. Control) using 2 sets of characteristics: (1) for the whole class and (2) just for the middle 14 minutes of the class, respectively. They used decision-tree algorithms to check for accuracy (99.3% with 1 model and 98.59% with the other), sensitivity (100% and 98.5%, respectively), and specificity (99.2% and 98.5%, respectively) in ADHD diagnoses. These findings indicate that a relatively short duration of fitness tracker or smartwatch wearing can be sufficient in most cases to detect ADHD and certainly that such an approach can be used as a reliable screener. This work builds on other actigraphy studies of children with ADHD that demonstrated discrimination between populations of children with and without ADHD via wearable data is reliable (Porrino et al., 1983; Halperin et al., 1993). More recently, in a study of 148 children (73 diagnosed with ADHD) wearable data resulted in 97.62% average sensitivity and 99.5% specificity using dominant hand actigraphs for 24 hours and convolutional neural networks for analysis (Amado-Caballero et al., 2020). Open questions in this space remain. Although researchers have attempted to identify ADHD sub-types using actigraphy (Dabkowska, Pracka, and Pracki, 2007), they have not yet been able to find differences between ADHD subtypes with movement data alone.

Using a more intensive approach, Muñoz-Organero et al. (2018) tested accelerometers, sensors for measuring acceleration and movement, on the dominant wrist and non-dominant ankle of 22 children (11 with ADHD, 6 of whom were also medicated) during school hours. They converted the sensor data into 2D acceleration images and trained a Convolutional Neural Network (CNN) to recognize the differences between non-medicated ADHD children and their paired controls. There were statistically significant differences in the way children with ADHD and those without moved for the wrist accelerometer (t-test p-value <0.05), but only between non-medicated children with ADHD and children without ADHD for the ankle accelerometer. This preliminary work indicates that such an approach might provide automated detection of ADHD with an accuracy of between 87.5% and 93.75%, high enough for a screener but not sufficient to indicate a truly diagnostic tool. After this study, Muños-Organero et al. (2019) used a Recurrent Neural Network (RNN) to improve their previous results, improving their screening tool in the process.

What is curious about these results is that a more intense period of data tracking than in Kam et al.'s work produced weaker results. As researchers move to solidify these results, and clinicians and educators move to put them into practice, tools must be tested with larger populations, and datasets should be open, combined, and compared.

Some researchers have gone beyond the use of accelerometry to include other sensor measures. For example, O'Mahony et al. (2014) used both accelerometers and gyroscopes, which together make inertial measurement units (IMUs), in a proof-of-concept study with 43 children, 24 with an ADHD diagnosis. They used a support vector machine learning approach, testing two algorithms: one with a linear kernel and one with a Gaussian kernel. The authors found that the Gaussian may have overfitted, and their general approach (i.e., the linear SVM) produced accuracy of 95.12%, sensitivity of 94.44%, and specificity of 95.65%, suggesting inertial sensors may be useful for diagnosing ADHD. In another IMU-based study, Ricci et al. (2019) measured linear and rotational movements of 37 school children, 17 with ADHD, finding similar rates of diagnosis with slightly different configurations of sensors. Kaneko, Yamashita, and Iramina (2016) also used acceleration and angular velocity sensors, this time on the backs of each hand and on the arm near the elbow. In a test of 33 children and 25 adults doing both an imitative motor task and a maximal effort task with one hand over 10 seconds, the researchers were able to quantifiably observe the expected development in pronation and supination based on age but could not distinguish between ADHD and typically developing individuals. The WEDA system, tested with 160 children ages 7–12, half with ADHD, attempted to discriminate between challenges in inattention from those related to hyperactivity and impulsivity, finding that the tasks cover all symptoms but perform better related to inattention (Jiang et al 2020). The scores of the overall tests discriminating between ADHD and typically developing children were highly sensitive and specific, but the system requires the child to wear 6 motion sensors (on the head, both hands, both feet, and the waist) while conducting 10 specific tasks, a setup that is unlikely to be used widely outside of research settings or specialized clinics.

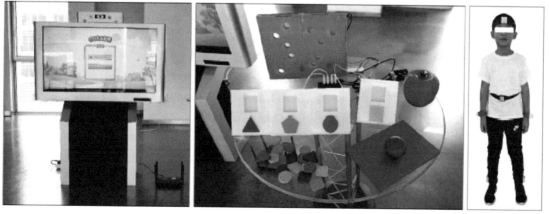

Figure 3.2: WeDA system including a touchscreen, 3D-printed physical devices, and motion sensors. Image from Jiang et al. (2020), used with permission.

Electro Interstitial Scan (EIS) is a galvanic skin response device that measures the concentration of free chloride ions in the interstitial fluid, the morphology of the interstitial fluid, and electrical stimulation (Maarek, 2012). Chua et al. (2019) sought to use this technology to augment other diagnostic work with children with ADHD. They collected EIS data from 182 Malaysian school children (58 with ADHD), finding that the system detected significant differences between the two groups. These results show promise as a complement to traditional diagnostic measures but have not been tested prospectively.

Moving to directly sensing the brain, in 2011, the ADHD-200 Global Competition was held to identify biomarkers of individuals with ADHD using resting-state functional magnetic resonance imaging (rs-fMRI) and structural MRI (s-MRI) from 973 individuals (Milham et al., 2012). Then, the Neuro Bureau prospectively collaborated with the competitors to preprocess the data and share their results at the Neuroimaging Informatics Tools and Resources Clearinghouse (NITRC) (http://www.nitrc.org/frs/?group_id=383). The repository was released and can be downloaded from NITRC without data usage agreement, but user registration is needed for non-commercial research purposes. From that time until now, several computer-based approaches have been explored, trying to identify either biological markers or patterns using the fMRI Benchmark. Eloyan et al. (2012) accomplish the best performance metric among the participants of the completion, but although it was the best, there is still necessary to develop efficient algorithms that can distinguish ADHD vs. neurotypicals. In the last five years, a variety of machine learning approaches have been used to classify MRI/fMRI/sMRI data from ADHD (Biswas et al., 2020). Machine learning fits well into behavioral science approaches as at its core, machine learning is about finding patterns in large-scale datasets, something quantitative behavioral scientists already use statistical methods to do. Machine learning, powered by extensive computational power, takes these methods to the next level. Once patterns are found, they can be used for a variety of diagnostic or intervention approaches.

Machine-learning approaches that have been used for ADHD imaging include support vector machines (SVM) (e.g., Sen et al., 2018; Riaz et al., 2018; Tan et al., 2017; Ghiassian et al., 2016; Rangarajan et al., 2014), Convolutional Neural Network (3D-CCN) (e.g., Zou et al., 2017; Ariyarathne et al., 2020; Wang et al., 2019), Extreme Learning Machine (ELM) (e.g., Peng et al., 2013; Sachnev, 2015), Deep Belief Networks (DBN) (e.g., Kuang and He, 2014), k-Nearest-Neighbor (Eslami et al., 2018), and multiple linear regressions (Miao et al., 2019). While these approaches are outside of the scope of this review, machine learning advances continue to provide new possibilities and should be monitored closely.

Important to consider for the future is the issue of comparing these approaches and actually determining what might work best, a staple of clinical and translational science outside of the broad areas of information technology. The style of computing articles, however, and the limited requirements and interest in shared datasets, make such comparisons extraordinarily difficult. For example, although these methods may use the same database, and even some of them the same machine learning technique, the way they use the data (i.e., either the subset of data used or how they partition the data to make the training and testing subgroups) makes it difficult to compare the results. It is impossible, based on published material, to determine which is best, and it may not even be possible with full replication was such replication possible with current approaches. Therefore, in the future, to make a fair comparison the models could use the same performance estimation method such as the same partition dataset method (e.g., k cross validation method) and same model evaluation metrics. Funders and journal editors should make a point to address this reproducibility crisis such that machine learning approaches can make safe, fair, and accurate advances alongside pharmacological, physiological, and behavioral practices.

The analysis of brain source localization of the EEG signal of individuals with ADHD has created new opportunities for the diagnosis and treatment of ADHD, starting more than 80 years ago with Jasper et al. (1938) reporting a slowing of the EEG rhythms at front-central sensors. This was one of the first indicators of a difference in the brain function of children described as hyperactive and impulsive. Yet, there are still several methodological limitations from a clinical and technological point of view (Cortese and Castellanos, 2012; Loo and Makeig., 2012).

Most of the research on the diagnosis of ADHD using EEG can be based on temporal, spectral, and spatial features of the EEG signals. With the raw EEG data, the spectral components can be obtained to compute the background state of brain activity. Alternatively, the EEG data can be segmented around an event (Lenartowicz and Loo, 2014). Thus, the EEG data is a combination of temporal, spectral special features that can be used individually or combined to assist diagnosis. Several approaches are proposed based on EEG signal analysis in the literature for diagnosing ADHD using machine learning approaches (e.g., Tenev et al., 2014; Muller et al., 2010), including neural networks (Mohammadi et al., 2016), nonlinear analysis (Khoshnoud et al., 2018; Boroujeni et al., 2019; Khoshnoud et al., 2015), decision support algorithms (Abibullaev and An, 2012), and autoregressive models (e.g., Marcano et al., 2016). As with the approaches for MRI/fMRI/sMRI data, these particular algorithms are outside of the scope of this review but should be followed and integrated into clinical practice as possible. And as with the machine learning challenges described above, replication and meta-analyses are likewise impossibly with existing analysis and publication policies and norms.

Taken together, this research shows a substantial and growing interest in studying the brain activity of people with ADHD. Unfortunately, brain activity is exceptionally complex, and to date there is no conclusive diagnostic tool based on brain activity analysis.

3.3   CHARACTERIZING ADHD THROUGH TECHNOLOGICAL INTERACTIONS

Although not strictly diagnostic in nature, in some cases, researchers were able to characterize ADHD or better describe and understand it through technological assessments. These projects speak to the potential for better discrimination among subtypes of ADHD, ongoing assessment of progress related to therapeutic interventions, and the potential for automated responses based on these detailed assessments.

Johnson et al. (2010) explored how accurate children with ADHD are when using a pen on a digitizing tablet, finding that children with ADHD showed difficulties in movement accuracy on the right side. These results suggest that there may exist a subtle spatial bias toward the right, which could be adjusted using assistive technologies (ATs).

In work focused on understanding the impact of medication on posture, Sarafpour et al. (2018) found that a mobile force platform to evaluate postural performance could accurately assess patient balance. They were able to identify changes in the motor performance related to balance among children with ADHD (38 children with ADHD, 23 who had current medication-based treatments, and 25 children without ADHD), finding that those taking medication had similar readings to those without ADHD and significantly different ratings from those with ADHD who were not taking medication.

Similarly, Chatthong, Khemthong, and Wongsawat (2020) sought to measure the impact of occupational therapy on school-aged children with ADHD. In a study of 305 school-aged children in Thailand, they used brain mapping performance to measure differences between children with and without ADHD, finding that the group with ADHD had higher emotional awareness and language comprehension than the group without ADHD.

3.4   CONCLUSIONS AND FUTURE DIRECTIONS

The research published to date reflects substantial and innovative efforts to attempt to make diagnosis and assessment easier, more scalable, and more reliable through technological means. While a wide variety of tools have been deployed against these challenges, many of them are making minor tweaks to an already relatively proven technological approach. Accuracy levels are low in some cases, but quite high in most, indicating strong promise for using such tools in both initial diagnoses and long-term symptom monitoring. However, none of these approaches have been tested prospectively at the scale required to make them truly diagnostic instruments. We suggest that in the case in which an approach is already relatively successful (e.g., with the use of accelerometers and other sensors), researchers should work with clinicians for large-scale validation rather than seek to tweak their algorithms and sensors more for marginal improvements in the accuracy. Ultimately, a tool is unlikely to be adopted and implemented in clinical settings without a strong evidence base for its utility.

Also promising is the notion of embedding assessment into the everyday lives of children and adults with ADHD. By incorporating assessment tracking into things like daily work applications, gaming, or unobtrusive sensors, we may be able to study the effects of a wide variety of contextual and environmental triggers on ADHD, understand how ADHD evolves over a lifetime, and provide feedback to people with ADHD who may then be empowered to evaluate which coping strategies are working well for them, experiment with new interventions, and generally take more control of their experiences.

CHAPTER 4

Attention and Other Cognitive Processes

Although ADHD is commonly thought of as relating to challenges with attention and other cognitive processes in both children and adults, there is no single profile for these challenges. People with ADHD have highly individualized experiences with their symptoms, at times, exacerbated or accommodated by the context of the world around them. Although ADHD is highly heterogeneous, most people with ADHD describe similar challenges with attention and other cognitive processes; concerns regarding self-regulation, self-control, and executive functioning in many ways underly all other related experiences and symptoms but in differentiated ways. Thus, in practice, clinicians, parents, teachers, and people with ADHD themselves often prefer a battery of psychological testing to assist with the diagnosis of ADHD as well as the profiling of the specific cognitive needs of the person with ADHD. This is a particularly complex set of issues and procedures, far out of the scope of a book focused on the intersection of technology research and ADHD. However, in this chapter, we first provide some brief, but essential background on the cognitive elements of the ADHD experience. We then discuss some of the ways in which technology has been applied to support cognition in individuals with ADHD.

4.1 SELF-REGULATION, SELF-CONTROL, AND EXECUTIVE FUNCTIONING

Self-regulation, self-control, and executive functioning all are umbrella constructs (e.g., Moffitt et al., 2011; Zhou, Chen, and Main, 2012) that encompass processes involved in exerting control over cognitive and behavioral processes. Davidson et al. (2006) described self-regulation as an ability that reflects mature cognition from a developmental standpoint:

> Mature cognition is characterized by abilities that include being able: (a) to hold information in mind, including complicated representational structures, to mentally manipulate that information and to act on the basis of it, (b) to act on the basis of choice rather than impulse, exercising self-control (or self-regulation) by resisting inappropriate behaviors and responding appropriately, and (c) to quickly and flexibly adapt behavior to changing situations. These abilities are referred to respectively as working memory, inhibition, and cognitive flexibility. Together they are key components of both "cognitive control" and "executive functions..." (p. 2037).

ADHD is characterized by difficulties in these cognitive abilities, although profiles for specific abilities may vary across individuals with ADHD. Generally, there has been a strong interest in how to improve these cognitive skills—attention, working memory, inhibition, and cognitive flexibility—using a variety of treatment methods. Some cognitive training programs have targeted a specific ability (such as focusing and sustaining attention), while others have targeted multiple abilities simultaneously (e.g., attention, working memory, inhibition). A large body of research has examined how to improve these skills with different treatment approaches. We do not address pharmaceutical treatments or other therapeutic programs here, as our focus is solely on approaches that have applied technology in treatment.

A current systematic review of digital health interventions for individuals with ADHD (Lakes et al., under review) identified more than 2,000 records that focused on ADHD treatment delivered via a range of technologies [e.g., web-based intervention for teachers of students with ADHD (Corkum et al., 2019); mobile devices (Davis et al., 2018); mixed-reality technology (Kim et al., 2020)]; and reported research conducted with individuals with ADHD; among these, the majority focused on cognitive training. These included 772 papers focused on cognitive training, including attention and working memory (WM) training, and 239 records addressing neurofeedback. Other less studied areas included serious games and virtual/augmented reality interventions. Given the focus of this book, in this chapter, we emphasize interactive technologies in support of ADHD and leave the reader to the systematic review for details outside that space.

4.2   COMPUTERIZED COGNITIVE TRAINING

Since not long after personal computers became widespread, computer interventions have been designed to improve attention, working memory, and other executive functions. In the marketplace and across the Internet, products claiming to remediate attention deficits or improve cognitive functioning abound and have generated considerable enthusiasm and hope for accessible computer-supported treatment options. Many commercial products are available on the market, advertising clinical benefits to individuals with ADHD and their families. The efficacy of these products, however, is largely unknown. Recent news reports provide a relevant example. Lumosity (https://www.lumosity.com/en/) markets computer games ("personalized brain training"), designed to improve cognition, reported in 2015 that it had reached 70 million members. In January 2016, Lumo Labs, the company behind Lumosity, was reported to have agreed to pay a 50-million-dollar settlement to the Federal Trade Commission in response to claims regarding false advertising. The commission found that Lumosity made claims about cognitive benefits without scientific evidence to support those claims. The company was ordered not to make further claims of benefits for the disorders it had targeted, including ADHD, without first gathering sufficient scientific evidence (Rusk, 2016).

Thus, although these products may be exciting and appealing, it is important to consider the current evidence base, some of which we review in this section.

Rapport et al. (2013) conducted a review and meta-analysis of 25 programs designed to train working memory, attention, and executive functions in children with ADHD. At the time, they concluded that training, short-term memory appeared to produce a modest improvement in short-term memory, but that training mixed executive functions or attention failed to significantly improve those domains. Rapport et al. also pointed out that the positive effects of interventions did not appear to generalize to real-life improvements in academic, behavioral, or cognitive functioning. In other words, evidence for "far transfer effects" (e.g., a positive impact on a measure of functioning in real life versus performance on a trained task) was "nonsignificant or negligible." As a result, Rapport et al. concluded at the time that "Collectively, meta-analytic results indicate that claims regarding the academic, behavioral, and cognitive benefits associated with extant cognitive training programs are unsupported in ADHD" (p. 1237).

In the same year, Melby-Lervåg and Hulme (2013) reported results of a meta-analysis of 87 publications focused on working memory (WM) training and concluded, "working memory training programs appear to produce short-term, specific training effects that do not generalize to measures of 'real-world' cognitive skills." Two years later, Cortese et al. (2015) published a review and meta-analysis of 16 randomized, controlled trials examining cognitive training outcomes in children with ADHD. These authors reported that there were significant improvements in total ADHD symptoms, inattention, laboratory working memory tests, and parent ratings of executive functions; however, they pointed out that when outcome measures were completed by blinded raters, effects were weaker, suggesting potential expectancy effects. They, too, concluded that the evidence "provided little support for cognitive training as a front-line ADHD treatment" (p. 171) but stated that it may have a role as adjunctive treatment to treat certain *neuropsychological* impairments.

In the most up-to-date and extensive review of working memory training, Novick et al. (2020) concluded:

> "Unfortunately, the evidence now strongly, if not decisively, indicates that WM training does not transfer to performance on nontrained tasks. The critics were right: WM training does not lead to long-lasting generalizable improvements in cognitive functioning. We're sure that this conclusion will be disappointing to many readers of this book, as well as to the many individuals who have purchased commercial brain-training products. We're sorry to offer such a gloomy outlook on the state of the art, but sometimes science leads us where it leads us" (p. 542).

Thus, recent reports from systematic reviews and meta-analyses have produced broadly consistent findings—the impact of cognitive training on ADHD symptoms and performance on nontrained tasks appears to be minimal. Although improvements are usually noted in the trained tasks (e.g., tasks involving computer testing of cognitive functions) and tasks similar to trained tasks, there is a lack of compelling evidence indicating that these programs produce improvement in cognitive functioning in other settings (school, work, home). Some researchers (e.g., Rapport et al., 2013; Cortese et al., 2015) conceded that the evidence may be restricted due to methodological limitations in the studies, but they and other clinical scientists have concluded that these products cannot yet claim to be front-line treatments for ADHD.

However, research in this area continues to grow and is likely to continue to do so, using both more advanced technological approaches to cognitive training as well as larger and more rigorous clinical trials. In one of the most recent controlled trials, Kollins et al. (2020) randomized 348 children with ADHD to intervention (the STARS-ADHD digital intervention) or a digital control condition. While they reported significant improvements on the primary outcome measure (performance on a computerized test of attention), they reported no significant improvements in ADHD symptoms or functional impairment; thus, their findings were consistent with prior research, demonstrating that cognitive training programs may promote "near transfer" of training effects (i.e., improvement in tasks similar to the trained tasks), but not "far transfer" (e.g., improvements in clinical symptoms or school behaviors). However, given their rigorous trial design and demonstrated impact on attention skills, this product, marketed as EndeavorRx, recently obtained approval from the United States Food and Drug Administration (FDA) as a video game prescription treatment for attention in children with ADHD (https://www.statnews.com/2020/06/15/fda-akili-adhd-endeavorrx/). This work is likely to inspire future research to develop and study such products. As brain-training approaches have not yet delivered compelling evidence of desired improvements in ADHD symptoms in real-life settings, future work should examine how to gauge the impact of training attention on real-life outcomes, which would clarify our understanding of the potential clinical significance of this program and others like it.

4.3   NEUROFEEDBACK INTERVENTIONS

Another research area that has received considerable attention involves using computer programs to evaluate brain activity with an electroencephalogram (EEG) and provide direct feedback to patients. Neurofeedback interventions are intended to help patients learn to regulate their brain activity in order to retrain underlying neural mechanisms involved in cognition and behavior (e.g., Enriquez-Geppert et al., 2019). In a review, Arns et al. (2014) discussed the history of attempts to use neurofeedback to treat ADHD and described diverse results, concluding that specificity is needed in the EEG parameters that are trained in order for interventions to be effective. They described three "standard neurofeedback protocols" that are the basis for their subsequent review (Arns et al., 2020) of the state of the evidence for neurofeedback interventions.

Arns et al. (2020) noted that a neurofeedback treatment regimen typically involves 30–40 sessions of training, and as mentioned above, there are standard protocols with specific targets that have some demonstrated efficacy. Providing a concise review of the current state of this field, Arns et al. (2020) summarized results of prior meta-analyses (Van Doren et al., 2019; Cortese et al., 2016) as well as individual trials and concluded that neurofeedback training using "standard neurofeedback protocols" appears to yield positive and sustained effects on parent and teacher ratings of ADHD symptoms. They concluded that neurofeedback protocols for the treatment of ADHD "can be concluded to be a well-established treatment, or 'efficacious and specific' in line with the APA guidelines" (p. 45). This was consistent with the conclusions of Enriquez-Geppert and colleagues (2019) who described neurofeedback as "a viable treatment alternative."

Emerging research has described the development of video games that incorporate some form of neurofeedback. For example, in a pilot study, Blandon et al. (2016) described a videogame customized for neurofeedback (Harvest Challenge) that was developed to use children's measured attention levels to control the videogame. Other neurofeedback videogame interventions have been compared directly to computerized cognitive training programs that also include a video game approach. In a pilot trial, Steiner et al. (2014a, 2014b) compared standard computerized attention training to neurofeedback. In the neurofeedback condition, children (ages 7–11 years) wore a bike helmet with embedded EEG sensors while playing a computer game involving flying an airplane. They were told that when they concentrated, the airplane would ascend and that if they did not concentrate, the airplane would descend. The standard computer attention training program (Brain-Train) included attention and working memory modules. Outcome measures were administered to parents, teachers, and children themselves. Children in both groups were compared to a wait-list control group; when compared to the wait-list control group, children in the neurofeedback group demonstrated improvements in parent-rated ADHD symptoms. No significant differences for either group were detected using teacher reports. Children in the standard attention training condition self-reported improvements in attention, suggesting that they perceived their attention as improved following attention training. This result is interesting but should be interpreted with caution given the potential for expectancy effects and the lack of corresponding evidence from other raters (teachers, parents).

4.4   VIRTUAL REALITY INTERVENTIONS AND SERIOUS GAMES BEYOND NEUROFEEDBACK

A number of publications in recent years have described the development of VR interventions and serious games to treat symptoms of ADHD. Many described projects in the early stages of development with user feedback that had not yet been subjected to a randomized intervention study. For example, Avila-Pesantez et al. (2018) described the design of an augmented reality serious game designed to improve attention.

One randomized intervention study compared a virtual classroom cognitive remediation program to either methylphenidate or a psychotherapy placebo condition (Bioulac et al., 2018). Among 51 children with ADHD (ages 7–11 years old), significant improvements were reported indicating that children in the virtual classroom condition demonstrated significant improvement on virtual classroom tasks and performance on a computerized test of attention and that improvement in these two areas was equivalent to improvements observed in the methylphenidate group. The methylphenidate group also showed significant improvement in ADHD symptoms.

The REEFOCUS ADHD management gaming system was defined to train delay aversion, inhibitory control, sustained and selective attention, motor coordination, and working memory (Kanellos et al., 2019). REEFOCUS can be played at home on a smartphone or tablet device or in a mixed reality mode in a clinical setting. The mixed reality mode has both tangible and augmented reality elements. Parents, teachers, and clinicians can monitor engagement with the REEFOCUS game, and the game is adapted both automatically in response to the child's progress and response to caregiver input. In a study with 75 children in Barcelona, the children who played the game found REEFOCUS to be understandable and learnable, while most of them liked the story and enjoyed playing the game. Parents mostly approved of the game. Although these results show promise, the game's effect on the areas it was intended to treat was not reported.

Although exergaming is discussed in more detail in Chapter 9 dedicated to motor behavioral and physical access, it is worth considering briefly in relation to its potential impact on cognition specifically. Exergaming has been used to simultaneously provide cognitive training and physical training. In several studies, one research group (Benzing and Schmidt, 2017, 2019; Benzing, Chang, and Schmidt, 2018) described how exergaming could serve as a support for executive functions, including inhibition and switching. The exergaming intervention reported in their research required attention, inhibition, switching, and speed of action. Other promising research has supported the notion that physically and cognitively active videogames can have a positive impact on symptoms in children with ADHD. In a randomized trial with 73 children (ages 6–13 years) comparing a full-body-driven videogame to active control (an age-appropriate videogame), Weerdmeester et al. (2016) reported that the full-body-driven videogame was associated with reductions in teacher-reported ADHD symptoms. Notably, the teachers were blind to experimental conditions in this study. Shema-Shiratzky et al. (2019) also demonstrated promise with an intervention that involved avoiding virtual obstacles while walking on a treadmill. While not technically an exergame, this approach shares commonalities with those of exergames, and in this case, was associated with improvement in parent reports of children's social interactions and problematic behaviors. Long-term positive effects were maintained in both memory and executive function after system use.

Wearable technology has also been explored to provide real-time assistance regaining attention for children with ADHD. For example, CASTT (Child Activity Sensing and Training Tool) uses a heart rate band, accelerometers on the arms and feet to sense movement, an EEG device, and a smartphone. A preliminary evaluation with 20 children, with and without ADHD, found that although using multiple wearable sensors was uncomfortable for children, and sometimes the notifications were unnoticed, monitoring physical and physiological activities in real time could potentially assist them (Sonne et al., 2015).

As our lives are increasingly lived in hybrid virtual/digital and physical spaces, and entertainment is increasingly related to gaming and media consumption, engagement with these as tools for therapeutic effect will remain widely appealing. Likewise, people increasingly have high-powered computation in their homes in the form of gaming systems, which are often more powerful than the laptops or Chromebooks used for educational and work purposes. Leveraging these platforms to extend therapies will be essential for greater access and higher engagement with interventions.

4.5   CONCLUSIONS AND FUTURE DIRECTIONS

Computer games or digital programs to train attention, working memory, and other executive functions have been a popular area of research in psychiatry, psychology, education, and other spaces related to ADHD. There are hundreds of studies describing such programs and their outcomes; however, the underlying technologies are by and large the same for most of these studies. The evidence to date indicates that cognitive training programs often can help improve performance on trained tasks or similar types of computerized tasks, but that benefits do not appear to generalize nor extend to performance in other settings. Some researchers have suggested that these tools may be a useful adjunct to treatment in some cases (Cortese et al., 2015), but others have drawn conclusions that warrant caution in the prescription of such programs (e.g., Novick et al., 2020). For developers and researchers interested in further pursuing technologically delivered cognitive training programs, work is needed to address how to enhance programs in a way that enables training to produce improvements that will generalize to real-life settings. It may also be the case, and is worth considering as a research community, that no technological approach to cognitive training has—or will be able to have—long-term effects that transfer out of the specific tasks or contexts of the cognitive training program.

On the other hand, scientific reviews of neurofeedback have been more positive, with Arns et al. (2020) concluding that neurofeedback protocols for the treatment of ADHD "can be concluded to be a well-established treatment, or 'efficacious and specific' in line with the APA guidelines" (p. 45). However, Enriquez-Geppert et al. (2019) noted that while neurofeedback may be a useful adjunct to treatment, in community settings there is a need for standardization of treatment protocols. Thus, future work in this space should address how technology can be designed in a way that enables practitioners to readily and consistently deliver evidence-based approaches to neurofeedback intervention in their community settings.

Although more nascent as areas, serious games, exergaming, and mobile applications are being applied in this space, there is room for much more work in these areas as technologies improve and become more accessible for individuals with ADHD. This work deserves further inquiry given the potential to support cognition using tools that are readily available and accessible through devices people already own. However, randomized trials with blinded assessments are needed to carefully evaluate the efficacy of products. Much has been learned in the past decade about how to conduct a controlled study of this nature; moving forward, these lessons should be applied in study designs, with careful attention to the study of generalizable effects to demonstrate that improvements detected using study measures actually translate into meaningful clinical outcomes. Engagement with gaming companies, or at the very least the opening of their platforms further to researchers, would accelerate this kind of development.

CHAPTER 5

Social and Emotional Development

In this chapter, we overview the technological approaches for supporting social and emotional skills for people with ADHD. Technologies have been used to support training and therapeutic interventions as well as real-time or *in situ* support.

There is a shift in the clinical, educational, and parenting approaches to supporting social skills in particular right now, with a greater emphasis on helping people with ADHD cope with environments that have not always been constructed with them in mind, rather than trying to force people with ADHD to change themselves as the default position. However, a wide variety of research in this space predates this shift. Notably, greater acceptance in schools, workplaces, and communities of differences in social engagement will likely change the approaches taken, both in technology-based interventions and those that do not use technology.

5.1 SOCIAL AND EMOTIONAL DEVELOPMENT AND ADHD

Social and emotional skills enable children to experience, regulate, and express their emotions in socially and culturally "appropriate" ways (Yates et al., 2008; Halle and Darling-Churchill, 2016). These skills help children get along with others, build relationships with peers and adults, and support skills such as cooperation, following directions, regulating emotions, and paying attention (Halle and Darling-Churchill, 2016). Social and emotional skills provide the foundation for children to build healthier relationships, solve problems effectively, cope with life challenges (Parlakian, 2003), and attain academic achievement (Blair and Diamond, 2008; Konold and Pianta, 2005).

Unfortunately, social and emotional difficulties are common among individuals with ADHD (e.g., Classi et al., 2012). Socially, children with ADHD often struggle to make friends (e.g., McQuade and Hoza, 2008) as they may have difficulty reading social cues and respecting the personal space of other children and may tend to interrupt others or change topics in the middle of conversations (Uekermann et al., 2010). These and other social skills are dependent on attention and impulse control. Impulsivity can contribute to a tendency to say or do something before thinking it through, which can exacerbate difficult social situations. Emotionally, individuals with ADHD often have difficulty tolerating frustration and regulating their emotions when facing a challenging or otherwise stressful situation. Emotional difficulties associated with ADHD include poor emotion regulation, aggression, and reduced empathy (Barkley, 2006; Anastopoulous et al., 2011).

Research has indicated that social difficulties are partially attributable to dysregulation of emotion (Bunford et al., 2018). Stressful circumstances may deplete our resources for self-regulation, such that any of us displays poorer self-regulation after an especially stressful day. This is demonstrated in how adults may be terse or abrupt with family members after returning home from a stressful day at work. Given their underlying difficulties with self-regulation, children with ADHD may be more vulnerable to stress, pressure, and fatigue than their neurotypical peers, leading to poorer self-regulation and higher rates of aggressive behaviors and rule-breaking (Erhard and Hinshaw, 1994; Hoza, 2007, 2005). Moreover, difficulties with emotion regulation coupled with pervasive social challenges and daily stressors, can contribute to the development of comorbid mood disorders. It has been estimated that up to 50% of children with ADHD exhibit depressive and anxiety disorders (Gillberg et al., 2004; Elia et al., 2008), adversely impacting their education, quality of life, healthcare, and wellness (Strine et al., 2006).

5.2   SOCIAL SKILLS TRAINING

Social and Emotional Learning is a subfield of educational psychology (Weissberg et al. 2015; Mahoney, Drulak, and Weissberh 2018). Social skills do not develop naturally without any human intervention in anyone. Children are coached to develop social skills by peers, siblings, parents, other family members, and teachers. Thus, social and emotional development is an area of substantial interest to both child development specialists and to those particularly focused on learning and education in childhood. It has gained some traction in the human-computer interaction space in the absence of ADHD (e.g., Littlefield et al., 2017; Slovak, Gilad-Bachrach, and Fitzpatrick 2015; Slovak and Fitzpatrick 2015; Slovak et al., 2016; Tanaka et al., 2016). Taken together, this research indicates that technology can be used effectively for social and emotional development, but as in the ADHD focused papers we describe below, additional research is needed to demonstrate more extensible outcomes from these promising practices.

One evidence-based non-pharmacological approach to supporting the development of social-emotional skills in individuals with ADHD is social skills training. Social skills training programs focus on teaching problem-solving, emotion regulation, impulse control, and verbal and non-verbal communication (e.g., Storebø et al., 2019). For example, training may help children learn to "read" common social cues, wait for their turns, recognize when the topic shifts in a conversation, identify emotions in themselves and others (Fohlmann, 2009), learn social norms, and understand societal "rules" (Liberman, 1988). Training programs go by various names, such as social skills training, child life, attention skills treatment, behavioral therapy, social skills treatment, psychosocial treatment, and cognitive behavioral therapy. These names of interventions also characterize a wide range of treatments for issues other than ADHD. For example, Cognitive Behavioral Therapy is one of the most extensively researched forms of psychotherapy (Butler et al., 2006; Rothbaum et al., 2000) that emphasizes changing thinking patterns in the person engaging in the therapy. Collectively, we use the term "social skills training" as a broad descriptor for programs aiming to support the development of social and emotional skills.

Social skills training for children with ADHD tends to involve a variety of supports and are frequently adapted and personalized by an individual teacher or social skills therapist or coach. Thus, we do not provide extensive details of these approaches here. However, to help with understanding the potential landscape for technological design, we overview two popular approaches to social skills training: role-playing and scripts. Role-playing typically involves defining the social problem in enough detail to reenact or adapt it, acknowledging the negative feelings, discussing alternate ways to respond, and finally practicing those alternate options by reenacting the negative interaction. You do not have to have experienced a situation to practice social skills, however, and so another popular approach involves generating and practicing scripts. In this case, rather than reenact a negative experience, a person or a group might imagine a potential social situation. In response to this imagined scenario, the group would then discuss potential approaches and likely outcomes, sometimes going so far as to physically write a social script in response to the potential scenario, such as in a collaborative classroom (Heemskerk et al., 2011). All people have sets of social scripts ready at hand, and making them explicit is a way to simplify and make clear these scripts for people with ADHD (Bye and Jussim, 1993) who might otherwise struggle to respond in the moment (Bickett and Milch, 1987; Hubbard and Newcomb, 1991). Such scripts can be generated using computationally-enhanced tools (Boujarwah et al., 2011), and other technological approaches to social skills training have also been examined, as detailed in the following section.

5.3   TECHNOLOGIES FOR SUPPORTING SOCIAL AND EMOTIONAL DEVELOPMENT

Technological applications have been developed with the goal of improving motivation, engagement, and social interactions. For example, robots have been used to mimic social interactions and provide opportunities for role-playing in a fun and non-threatening environment. Also, serious games have been proposed to teach children social strategies, practice collaborative behaviors, and follow societal rules, such as practicing taking turns or engaging in teamwork. Online coaching also has been used to provide support and guidance through daily activities that may trigger emotional challenges. In the following paragraphs, we explore some of the research in these and other areas.

Serious Games are interactive media experiences and games that support people in learning new skills, thinking about difficult concepts, or accomplishing a variety of tasks (Michael and Chen, 2005). These games can be used for a variety of supports for people with ADHD. For example, Bul et al. (2015) developed and studied "Plan-It Commander," an Internet-based treatment for children ages 8–12 years who have ADHD. Building on self-regulation theory (Barkley, 2006; Cameron and Leventhal, 1995), Social Cognitive Theory (Bandura, 1986), and Learning Theory (Kato et al.,

2008) as conceptual models, the game teaches time management, planning and organizing, and prosocial skills and can be played at home independently by children. The player takes on the role of a space captain, gathering rare minerals throughout the universe for an interplanetary organization. Individual assignments within the missions ask children to solve problems that specifically engage the issues that are the subject of the intervention (time management, planning and organizing, and social skills). The game included a social community in which children could interact with one another. The game dynamic consists of three minigames: a labyrinth to learn how to manage time; Explorobot, to learn planning; and space travel trainer, to learn to help their team and to develop prosocial behavior. Bul et al. (2016) conducted a 20-week randomized controlled trial (RCT) with 182 children (ages 8–10 years) with ADHD from 4 mental health care clinics. The experimental group played with the serious game and continued treatment as usual for 10 weeks. After that, the experimental group continued with just treatment for another ten weeks. The control group received the same, but in reverse: first treatment as usual, and then treatment as usual with the serious game. The serious game ("Plan-It Commander") was played for up to 65 minutes per day up to 3 times per week. Parents and teachers reported improvement in social skills surrounding gameplay but planning and organizing skills were not significant between groups at either time period. The authors note that future work should include more cooperative gameplay to improve social benefits (Bul et al., 2016, 2018). Cooperative gameplay has been shown to have positive effects on social skills for children with autism (e.g., Boyd et al., 2015; Piper et al., 2006). However, no similar studies were found for children with ADHD.

Research examining the benefit of serious games, such Plan-It Commander, suggests that children with ADHD can learn and practice social skills in a virtual environment and that they can be empowered to practice those skills with their peers using technological tools. On the one hand, virtual environments create an immersive system where researchers or therapists can personalize stimuli and facilitate interaction between children. On the other hand, tools that facilitate face-to-face interactions may provide more ecological validity for developing certain social skills, but these face-to-face interactions can cause anxiety in some children with ADHD and may be more difficult to control. This trade-off between virtual socialization and face-to-face social engagement should be explored further. Specifically, more research is needed to understand how skills developed in a virtual environment can be transferred to support face-to-face social interactions.

In contrast to intervening in a virtual environment, other researchers have explored how to support face-to-face social interactions, such as engagement with a therapist and participation in therapy, using technology-based solutions. One therapy that has targeted socio-communicative behaviors for children with neurodevelopmental disorders, including children with ADHD, is music therapy (Crowe and Rio, 2004). In order to improve interactions between children and a therapist and to increase attention during music therapy sessions, Lobo et al. (2019) co-designed with therapists CHIMELIGHT, an Internet of Things (IoT) music therapy tool. CHIME-

LIGHT was designed to be used by children and their therapists during a hand chime activity. Each device contained a microcontroller, Bluetooth, RGB LEDs, accelerometer, gyroscope, and magnetometer. CHIMELIGHT monitors performance and provides visual rewards (e.g., static blue light when the child is still and a filling yellow light when the child performs a hand chime movement successfully). Therapists can add other feedback through a mobile app as well as analyze assessment data. A case series study, using an ABBAAB design, was conducted to evaluate CHIMELIGHT. Three children with ADHD took music therapy sessions for six months using CHIMELIGHT with and without feedback. Quantitative results and observations indicated that when the device delivered visual feedback, children were more engaged (e.g., looked at the chime) and exhibited fewer negative behaviors (e.g., looking away, non-imitative behaviors). This preliminary yet promising research suggests that changes in socio-communicative behaviors, such as synchronized musical play and looking at the therapist, were changed by augmenting music therapy with technological-tools. Future work should look to scale such evaluation as well as to understand whether and how such indicators of engagement with the other people in the musical session translate to engagement in the social world or make such engagements less challenging and/or more pleasant for people with ADHD.

Face-to-face approaches to improving social skills have also incorporated robots. Researchers in the human-robot interaction space have explored how robots may help facilitate social and cognitive development for children with ADHD. For example, Lehmann et al. (2011) compared the effectiveness of a humanoid social robot against a mobile robotic platform in supporting social interaction through play scenarios. The humanoid robot, named Kaspar (Kinesics And Synchronization in Personal Assistant Robotics), was developed by the Adaptive Systems Research Group, University of Hertfordshire (Dautenhahn et al., 2009), predominantly for children with autism. Kaspar is a minimally expressive child-sized humanoid robot explicitly designed to promote communication and social skills in children with special needs such as autism and ADHD. Kaspar has a nose, eyes, and mouth and can move the head, arms, and face while interacting with people. Kaspar's face is a mask used in CPR training with eyes that open and close and some minimal ability to express emotions, including tilting of the head, smiling, and lowering lips to portray a sad face. Kapsar has tactile sensing capabilities built into its cheeks, torso, arms, back, palms of the hands, and soles of the feet. These sensing capabilities allow Kapsar to appear to respond to a child's touch, which can be used to encourage or discourage particular behaviors. Children do not report being unnerved by KASPAR's humanoid appearance.

Figure 5.1: A pair of Kaspars, child-like robots from the University of Hertfordshire, plays with a child during an imitation game. Image courtesy of Ben Robbins.

IROMEC (Marti et al., 2009; Marti, 2010) is a robotic platform developed as a social mediator for children with special needs. IROMEC has many tangible components that modify the appearance and behavior of the robot. The robot can move in space autonomously and is remotely controlled. It has a digital screen that displays graphical interface elements, like facial expressions. A within-subjects study was conducted with ten children (nine boys) with neurodevelopmental disorders, including ADHD. Children interacted with KASPAR or IROMEC in individual sessions. Each session consisted of introducing the robot, followed by interaction scenarios, and ending with time to say goodbye. Each session lasted 20 minutes. Children participated in three scenarios during each session. These scenarios aimed to support cause and effect, imitation, and turn-taking. For example, in the turn taking scenario, children seek to understand cause and effect through engagement with the robot.

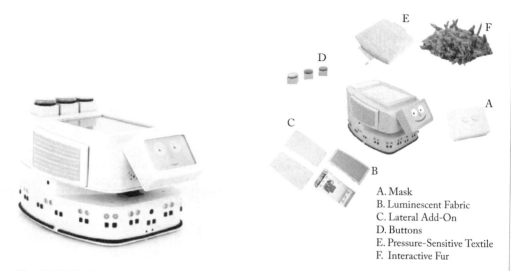

A. Mask
B. Luminescent Fabric
C. Lateral Add-On
D. Buttons
E. Pressure-Sensitive Textile
F. Interactive Fur

Figure 5.2: The IROMEC Robot; this nonhumanoid robot has a face displayed on its "head," which is a screen positioned to one side. In the image on the right, the authors describe the components, including a mask, luminescent fabric, buttons, pressure sensate textiles, and interactive fur. Images courtesy of Patrizia Marti.

In a study of children with autism using both robots, results indicated that imitation was better supported with Kasper (the humanoid robot), as children were more willing to imitate the robot's movements (Iacono et al., 2011). However, this type of comparison has not been conducted with children with ADHD.

Collectively, this research shows that technological interventions have the potential to support social skills interventions for children with ADHD, by providing opportunities to practice behaviors and receive feedback in a safe environment. Technologically supported environments have the potential, although are not guaranteed to provide such safe environments. For example, for children who face barriers in access to traditional therapies, at home therapies enabled by games, robotics, and so on can provide an important resource. Virtual simulations and serious games can provide social engagement and skill building without any physical interaction. Robots can provide some physical engagement, especially in cases in which therapists individualize the robot-supported intervention by controlling stimuli and feedback. VR systems, particularly those that do not require a full room, may be more readily accessible and affordable to families than robots, but both should be explored in larger studies, given the likelihood of the technology to become more affordable and consumer grade over time. At the same time, as the move to fully remote engagement during the COVID-19 pandemic has reminded us, homes are not always the safest place for children. Some children, especially those with ADHD, are at greater risk for abuse in the home (Brisco-Smith and Hinshaw, 2006) and reliance on parents to implement or supervise home-based therapies may increase the already-high burden on parents who are facing substantial challenges associated with work, parenting, and other responsibilities. Thus, technological supports for monitoring child safety and minimizing parent burden should also be considered, and any solutions that remove outside engagement from teachers, therapists, and social workers in their entirety should be examined skeptically.

5.4   EMOTION REGULATION

Emotion regulation difficulties are a key mechanism for the risk of depression in young people with ADHD (Seymour et al., 2012; Steinberg and Drabick, 2015). Children and adolescents with ADHD often exhibit difficulty modulating the intensity of their emotions (both positive and negative) and controlling emotions in frustrating situations (Cole, Martin, and Dennis, 2004). Emotion regulation requires that an individual first recognize and monitor their emotional state, which requires awareness and attention. It then requires self-evaluation, wherein the person can evaluate this emotional state. If the state is undesirable (e.g., the person is experiencing sadness or anger), the next step is to address this state using a learned strategy. For example, therapies for depression often focus on teaching individuals to choose to engage in a pleasurable or mood-improving activity (e.g., going for a walk, listening to music, gardening, playing a game, reaching out to a loved one) when they begin to feel lethargic or depressed. Individuals with ADHD may have difficulty modulating and controlling emotions because of insufficient self-monitoring, self-evaluation, and self-correction skills. The development of these skills can be taught in therapy and could also be supported using innovative technological tools.

Ecological momentary assessment (EMA), in which people report their behavior or attitudes on a schedule or in response to a trigger (Shiffman, Stone, and Hufford, 2008), has been used as an approach to study and intervene to support emotion regulation in children with ADHD (e.g., Rosen and Factor, 2015; Wen et al., 2017). Although EMA began as an approach to studying people's behaviors and perceptions in real time, ADHD researchers and providers have taken this approach as an intervention as well as a data collection approach, mirroring to some degree the notion of personal health informatics (Dishman, Matthews, and Dunbar-Jacob, 2004). One study illustrated how self-monitoring can be achieved and may support emotion regulation in girls with ADHD (Heron, et al., 2019). The assessment of emotion can be done more frequently than might be possible using a self-report approach like EMA by using naturalistic real-time mobile technology (Babinsky and Welkie, 2019). In a pilot study, 13 adolescents with ADHD and depression (ages 12–16 years old) and their mothers were provided with cellphones that triggered surveys at random times during one week, without interfering with school time. Each phone survey captured negative emotions and their intensity using six items [i.e., lonely, sad, worried, bored, hopeless, and in pain (stomachaches, headaches, etc.)]. Each emotion was rated on a scale from 1 (very slightly or not at all) to 5 (very much). Both parents and girls rated their emotions on the phone. The study illustrated how mobile applications provide a feasible method of monitoring emotions in individuals with ADHD, and the significant correlation between child and parent scores provided some evidence for the validity of self-reported emotion in youth with ADHD (Babinsky and Welkie, 2019).

Web-based coaching also has been used to promote emotion regulation among individuals with ADHD. Coaching can provide effective strategies to help individuals respond to problems and regulate emotions in everyday life in an accessible and acceptable manner (Goldstein, 2005; Murphy et al., 2010; Parker and Boutelle, 2009). Researchers took a user-centered design approach to develop a web-based chat intervention called SalutChat (Sehlin et al., 2018). Adolescents and young adults with ADHD and/or autism and their parents were invited to design SalutChat to connect with psychologists and therapists as coaches. During a six-month pilot study with ten adolescents and young adults with autism and/or ADHD (15–26 years old), participants reported improvements in self-esteem and perceptions of quality of life. These findings suggested that web-based coaching may be a useful complement to other interventions for people with ADHD (Wentz et al., 2012). Moreover, caregivers' burdens decreased when the adolescents were using the web-based intervention (Söderqvist et al., 2017), an important consideration in any intervention aimed at helping youth with ADHD. Finally, after the conclusion of the pilot study, a qualitative follow-up study was conducted to examine the overall experience of adolescents using SalutChat. Results supported prior findings, indicating that web-based interventions could play a useful role in supporting adolescents and young adults with ADHD and autism. Although not a replacement for face-to-face therapies and other treatments, this approach could be a promising complement or alternative to other support and treatment options (Sehlin et al., 2018).

Technological tools to support emotion regulation for youth and adults with ADHD have the potential to have a meaningful impact on their quality of life and outcomes in a number of domains. These tools can help individuals monitor and evaluate their emotional states and prompt them to engage in behaviors that will help move them into more positive emotional states. As illustrated above, this could be accomplished using mobile applications as well as web-based coaching and support. In our work, we are exploring how smartwatches can be used as tools to support emotion regulation in children and adolescents with ADHD (Cibrian et al., 2020a). These mobile applications have the advantage of being readily accessible as well as being delivered through devices individuals already own and interact with on a daily basis. Given the growing technologies that enable researchers to collect a variety of contextual data in an individual's life using wearable devices (e.g., sleep patterns, physical activity patterns, noise in the environment), our future research aims to explore how this data could be used to predict potential situations that might deplete self-regulation such that an application could pre-emptively deliver interventions (e.g., prompts to engage in self-regulatory behaviors).

5.5 CONCLUSIONS AND FUTURE DIRECTIONS

Living with ADHD as a child, adolescent, and an adult can require developing a wide range of coping skills to deal with the social and emotional stress of the context in which you live, work, and study while adapting personal experiences with ADHD. Socially, children with ADHD can struggle to make and maintain friendships, while adults may enter or stay in unhealthy relationships as a result of impulsivity or other challenges. The stress associated with these unhealthy or absent relationships can interact with existing challenges around self-regulation to make for a social life characterized by anxiety and fatigue. These same impacts occur in educational contexts, the workplace, and within families. However, the research in technologies in this space shows much promise for future interventions and supports.

In this chapter, we outlined the ways in which a variety of technologies can be used to augment or replace human coaching and therapeutic interventions. From simple online telehealth solutions to more elaborate gaming and robotics approaches, technologies can support, personalize, and scale a wide variety of therapeutic approaches. For people whose lives are already very busy and for whom the need for additional therapies informs how their lives must be structured, support available at home, on demand, and/or customized to individual context can be invaluable. In particular, emotion regulation often requires support at exactly the time that a person with ADHD might most struggle to reach out. Although most of these technologies have not been tested at scale nor validated with the kind of clinical evidence we will ultimately need, there is promise for real-time, contextualized support.

An individual's lifetime trajectory of social and emotional development is inherently interactive and collaborative, informed by the norms, communication standards, and engagement of others in the communities in which individuals live and work. Thus, collaborative technologies must be examined more carefully in this space, with an eye both toward making it easier and more enjoyable for people with ADHD to interact within the dominant culture and toward making it easier and more enjoyable for those without ADHD to interact with people with ADHD, understand their perspectives, and support their full inclusion.

CHAPTER 6

Behavior Management and Self-Regulation

Behavioral interventions have been used to support the management of ADHD symptoms (hyperactivity, impulsiveness, and inattention) by teaching individuals with ADHD how to control their behaviors (Friedman and Pfiffner, 2020; Reid, Trouth, and Schartz, 2005). Self-management is important for all people, regardless of an ADHD diagnosis, and improved skills in these areas tend to result in long-term success in education, work, and other areas of life.

6.1 BEHAVIOR MANAGEMENT, SELF-REGULATION, AND ADHD

One common goal of behavioral interventions is to improve the control of attention and impulsivity through the development of self-regulation skills (Reid, Trouth, and Schartz, 2005). Self-regulation involves managing one's behavior, emotions, and thoughts to pursue long-term goals. Self-regulation requires self-monitoring, reflective thinking, goal setting, decision making, self-evaluation, and management of emotions arising as a result of behavior change (Murray and Rosanbalm, 2017; Reid, Trouth, and Schartz, 2005). As described in Chapter 4, self-regulation is critical to regulating emotions. Here we discuss how this construct is applied in understanding and intervening to improve behavior.

Self-regulation is fundamental to adaptive developmental tasks at all stages of life (McClelland et al., 2018). Self-regulation first develops in early childhood (Bronson, 2000; Kochanska, Coy, and Murray, 2001; McClelland and Cameron, 2012) and continues to develop throughout childhood and adolescence. Cultural context (Raver, 2004), parenting approaches (Bernier, Carlson, and Whipple, 2010), and poverty and stress (Evans and Kim, 2013; Raver, 2004) all can influence the development of self-regulation with challenges continuing into adulthood for many who struggle as children. As noted in our introduction, longitudinal research following children from birth into adulthood has documented the critical importance of self-regulation for a variety of health, economic, and educational outcomes (Moffit et al., 2011).

Despite the continuing concerns into adulthood of challenges with self-regulation, and difficulties with impulsivity and inattention in particular, we found no research in our review focused on using technology to support behavior management and self-regulation in adults with ADHD. Thus, this chapter focuses on research describing the use of technologies for children, most commonly school-aged children. To support children who struggle with self-regulation, caregivers (e.g., parents, teachers) often use motivational or emotional scaffolding. Motivational scaffolding involves the caregivers' efforts to initiate and sustain children's enthusiasm for a task, using praise and encouragement, redirection of the child's attention, or restarting the task (Gulsrud, Jahromi, and Kasari, 2010) [i.e., co-regulation (Ting and Weiss, 2017)]. When caregivers use co-regulation strategies successfully, children with ADHD are better able to reduce problematic behaviors and increase successful behaviors; these successes can lead to the development of greater feelings of personal self-efficacy and confidence and also may improve parent-child interactions (Danforth et al., 2006; Gisladottir and Svavarsdottir, 2017; Loren et al., 2015). Therefore, some technologic interventions have focused on supporting parents by teaching them behavioral intervention and co-regulation strategies.

To support behavior management and self-regulation skills for children with ADHD and their caregivers, researchers have explored the use of web and mobile-based interventions to provide training and support. Research has examined mobile technology to support tracking children's behaviors and strategic prompting. Serious games and robot-based technologies have delivered behavioral therapy, and sensor devices and EEG have been used to support bio- and neurofeedback training with promising clinical results. Technologies have been tested in laboratory, home, and school settings.

In this chapter, we overview the technological applications designed, developed, and tested to support behavior management for individuals with ADHD, behavioral therapies, and the development of self- and co-regulation skills (self-regulation of emotions is discussed in Chapter 5). In the following sections, we explain these different approaches, providing examples from recent literature.

6.2   TECHNOLOGY TO SUPPORT BEHAVIORAL TRAINING FOR CAREGIVERS

As a first step to understand behavior management and self-regulation, it is beneficial for caregivers and peers of individuals with ADHD to understand the challenges faced by individuals with ADHD with the hopes of increasing empathic responses. Toward this aim, Goldman et al. (2014) developed a videogame intended to recreate the experience of ADHD through game mechanics. The game, called Drawn to Distractions, is a third-person 3D videogame that aims to cause frustration and mental exhaustion in players (to mimic ADHD symptoms). Players need to help a child-character to reach the school library in 10 minutes. The video moves very slowly to look tedious, and there are bubbles and spaceships randomly appearing. If the players touch a spaceship, the character is transported in a colorful and engaging mini-game (to mimic a rewarding activity distracting from the main goal). The videogame was tested in three user studies: one included 7 caregivers, another included 13 undergraduate students who had friends with ADHD and the third included 8 undergraduate students and graduate students. The research indicated that it was feasible to use persuasive games to promote understanding of ADHD behaviors among caregivers and friends of individuals with ADHD. Approaches such as these are novel approaches to increasing empathy toward and understanding of individuals with ADHD, which could ultimately improve the life experiences of those living with ADHD; however, more work is needed both to study and disseminate such tools.

However, creating empathy and awareness alone is not enough, especially for caregivers who want to promote self-regulation and improve behaviors in their children. Although treatment to improve behaviors and self-regulation is highly recommended, fewer than half of families receive this type of support, largely because face-to-face therapy can be perceived as expensive, inaccessible, or unengaging (Chacko et al., 2016; McEwan et al., 2015; Kern et al., 2007). To address some of these barriers to treatment, researchers have explored how to used web-based technology to support parent behavioral training (Dupaul et al., 2018; Breider et al., 2019; Olthuis et al., 2018; Ryan et al., 2015). For example, DuPaul et al. (2018) conducted a randomized controlled trial comparing face to face behavioral training with online behavioral training. All families received ten sessions of parental behavioral training. Both interventions had high attendance and improved the parent's knowledge of behavioral training, suggesting that web-based interventions can produce similar clinical outcomes when compared to in-person interventions. This result is particularly reassuring given the recent and sudden shift to telehealth treatment in response to COVID-19 pandemic control measures. In early 2020, many mental health providers across the world (including one of this books' authors) rapidly transitioned from in-person therapies to telehealth therapies relying on the support of a handful of video-conferencing web-based platforms. We anticipate that reports examining telehealth treatment experiences and outcomes will increase rapidly in the next few years as a result of this recent shift and hope that they will advance telehealth research and treatment. Researchers and clinicians have long viewed telehealth as a promising means by which to reduce barriers in access to treatment (e.g., geographic distance from providers, time constraints), but barriers associated with health insurer reimbursement and other administrative factors had stifled the growth of telehealth in spite of rapid growth in technological capabilities. Anectodal reports in 2020 have indicated that patients and families view telehealth as convenient and effective, with patients (including many in our clinics) and providers indicating that they would like to see increased use of telehealth in the post-pandemic era.

In the last decade, technology research has shifted its focus from delivery of intervention through personal computers to mobile devices. Mobile devices have added benefits due to portability as well as embedded sensors that allow for tracking behaviors and physiological data. Mobile devices have added flexibility as they can be used for web- or app-based telehealth delivery as described above, but also can be used to support treatment throughout the day, outside of traditional appointments with providers. Mobile technologies may increasingly allow us to provide personalized intervention to children and their caregivers, delivering prompts in response to particular individual indicators that intervention is needed. Among parents, mobile intervention provided in tandem with wearables has shown the potential to reduce parental stress and increase adherence to therapy (Pina et al., 2014). For example, ParentGuardian involved a system consisting of a mobile phone, a peripheral display, and an electro-dermal wrist sensor that together gathered information about parental stress and provided strategies to support behavioral improvements in children. A two-week deployment study demonstrated that in situ cues can remind parents to implement strategies in the moments when they are most needed, but automatically detecting stress using the wrist band was still a work in progress at that time as it is now. This particular area, while largely unexplored to date, holds particular promise as technologies continue to improve, increasing the potential impact of such tools.

6.3   TRACKING, PROMPTING, AND REINFORCING BEHAVIORS

Technology is particularly promising when it comes to individualizing intervention, particularly by tracking data, either automatically collected using a sensor or through direct input from users. This tracking is then used to provide reinforcement or prompting to promote behavior change. This approach has been studied with both children with ADHD and their caregivers.

Among children, Aase and Sagvolden (2005, 2006) conducted a study examining the impact of reinforcement on behaviors. In a sample of 56 children with ADHD (from 6–12 years old) and 64 children without a diagnosis of ADHD, children were asked to play a very basic videogame on a computer and to try to figure out how the game works. The game consisted of two squares displayed on a computer screen. Every time the children clicked over the squares, they changed colors as feedback, but when they clicked in the correct one (one of the squares), they received a reinforcement every 2 seconds (high density) or 20 seconds (low density). Results showed that when reinforcement density was low (every 20 seconds), there were differences among children with and without ADHD, but when it was higher, there were no differences in activity, impulsiveness, or sustained attention. This research demonstrates how technology can track game-related behaviors and suggests that children with ADHD can be supported with automated reinforcement and that the timing of reinforcement may be important.

Another approach to tracking behaviors involves asking individuals with ADHD directly or through their caregivers to input data. For example, Wills and Mason (2014) developed a self-monitoring application to improve on-task behaviors. The application, called I-Connect, was piloted with two adolescents with ADHD (ages 14 and 15 years). I-Connect sent scheduled prompts to participants (e.g., "Are you on task?") and provided an opportunity for data input (e.g., press "yes" or "no"). After more than 20 sessions of use following an ABAB design methodology, researchers found that both students improved their on-task behavior, suggesting that self-reflection using prompting may increase awareness of one's behaviors.

However, for younger children, rating one's behavior may be more difficult, and self-monitoring may need to be learned in order to accurately assess one's behavior. For example, Schuck et al. (2016) develop the iSelfControl app, an iPad application, where every 30 minutes students with ADHD and their teachers were prompted to rate children's on-task behaviors. Twelve children (ages 9–11 years old) and one teacher participated in the study in a school setting. For 13 days, both students and teachers rated behaviors, including following directions, adhering to classroom rules, staying on task, and getting along with peers. Results indicated that students initially demonstrated a weak self-awareness (based on comparison of their self-ratings to ratings of their behavior provided by their teacher) that gradually improved over time, suggesting that self-monitoring can be learned and can improve with the use of technological supports.

In primary school, in addition to learning to regulate their behaviors, children learn important social skills, like collaboration, empathy, and awareness of others, which can be essential to long-term success in education, social interactions, and other key domains of daily life. Thus, to engage these collaborative skills, particularly, Matic et al. (2014) developed and evaluated the use of a shared classroom display in a school that was already using an intense individualized behavior management program. Each child's individual behaviors contributed to a portion of the shared display, with a rewarding image being shown at the end of the day in classrooms in which all children succeeded in their behavior management goals for the day. In a study with 28 students with ADHD (8–12 years) across multiple classrooms in the same school, students' behaviors improved during the intervention phase, but perhaps more importantly, students also began to support one another, encouraging each other to do better and congratulating each other on successes.

Figure 6.1: Paper prototypes resulting from the co-design work in Cibrian et al. (2020a). On the left, the smartwatch format used to engage the children with ADHD in design activities is shown. On the right are examples of the ways in which the children imagined what their interfaces might say to support them in social interactions (Daily Goal: Help a friend), social well-being, and health (Challenge: Be nice, Call Bro, Run for 20 minutes), and physical activities (walk, run, bike).

More recent research has explored how smartwatches can be used to support behavioral therapies. Cibrian et al. (2020a) described design guidelines to develop self-regulation applications for smartwatches. They asserted that using a hybrid approach to promote self- and co-regulation using a smartwatch for children and a paired smartphone application for parents may be an especially promising approach to supporting behavioral intervention in the home. They highlighted the advantages of smartwatches, such as their automatic tracking of behaviors, ability to deliver timely prompts, and the potential to deliver intervention discretely, using a mainstream device to avoid stigma among children and adolescents.

Some research has explored how to detect hyperactive behaviors in children with ADHD automatically with the aim of delivering timely prompts. For example, Shih (2011) and Shih et al. (2011, 2014) explored the use of input devices such as mouse and Nintendo Wii remote to detect high-performance limb action to trigger prompting to the student when standing up arbitrarily during class (a particular hyperactive behavior observed in children with ADHD). Three studies, one with a mouse and two with the Nintendo Wii remote, showed that these input devices are able to infer posture, and participants were able to maintain a static limb posture during the interventions. Overall, these research studies suggested that it is possible to infer behaviors of children in order to help them to be more regulated. However, in practice, approaches to helping children regulate hyperactivity should be balanced with accommodations that allow children to move when necessary, for example, by allowing frequent activity breaks.

In summary, technologies can be used to promote self-regulation by increasing self-monitoring (by tracking behaviors and providing a visual representation of behaviors), by supporting self-evaluation or self-reflection (by prompting individuals to consider whether or not their behavior is appropriate for the situation), and self-correction (by delivering prompts and/or specific strategies to promote positive behavior change).

6.4   TECHNOLOGY-BASED THERAPY

Technological support for mental health has been an area of interest for multiple decades (Lanyon, 1971). More recently, researchers have focused their efforts on approaches that leverage smartphones (e.g., Luxton et al., 2011; Kretzschmar et al., 2019), robotics (e.g., Fiske, Henningsen, and Buyx, 2019; Riek, 2016; Tamura et al., 2003), and intelligent agents (e.g., Kretzschmar et al., 2019; Vaidyam et al., 2019). These new technologies provide promise across a broad spectrum of mental health approaches, though in this section, we are focused on technologically-enabled therapeutic approaches for people with ADHD, specifically.

Particularly novel (and potentially controversial) in this area is the notion that robots or virtual agents can act as the mental health professional themselves, initially augmenting but perhaps one day replacing this role for a subset of mental health needs. The limited resources available for mental health support necessitate an approach that scales and technology-augmented therapy has that potential. For example, Clark-Turner and Begum (2017) explored how to use reinforcement learning to train a Nao-robot to deliver behavioral interventions for children with ADHD. Although it was not empirically validated for use with children, the robot was successfully trained using a model-free reinforcement learning algorithm (Q-learning) to create an intelligent model to teach the social greeting steps (greeting command, prompt, and reward). A pilot study was conducted with 11 students without ADHD to measure the success of the model when delivering positive or negative prompts, given the greeting command. The accuracy of the model was 83.3%, which indicates some potential for robot-led therapy that includes behavioral prompting.

Virtual environments also have been explored as tools to deliver therapy for individuals with ADHD, by focusing on behaviors perceived to be negative or counterproductive by others, such as impulsivity, and transforming them into positive behaviors, such as careful reasoned choices. For example, Weerdmeester et al. (2016) compared an exergame called Dragon to the exergame version of Angry Birds, examining the impact on teacher-rated ADHD symptoms. Dragon allowed children to embody a small dragon to save its world through three levels: (1) "the forest," focusing mostly on attention and impulsivity; (2) "the water tower," focusing on hyperactivity; and (3) "the cave," addressing impulsivity and motor skills. The game provided positive auditory feedback (e.g., "You're doing great!") for both successes and failures. After six 15-minute gameplay sessions, in a study with 73 children (ages 6–13 years) with symptoms or diagnosis of ADHD, children who played Dragon exhibited a marginally greater improvement in terms of teacher-reported ADHD symptoms, such as impulsivity, than those who played Angry Birds.

Perhaps surprisingly or even counterintuitively, videogames also have been used to address videogame addiction[1] through the "positive" use of technology. For example, in a demonstration study, Ruiz-Manrique et al. (2014) developed the Tajima Cognitive Method (TCM) as part of a mobile/tablet application with exercises to train attention, memory, calculation, visual-motor coordination, and perceptual reasoning. TCM is a cognitive training method that was designed to enhance working memory, reasoning, and attention. The application, called "ADHD Trained," was used to treat a 10-year-old boy's videogame addiction in tandem with pharmacological intervention. This study results suggested that videogames could, at least in this case, increase the motivation of children during therapy and reduce the behaviors perceived around their videogame use perceived to be negative by parents, friends, and providers, such as perseveration on the game, hyperactivity, and feelings of restlessness or irritability. As this was a single case study, far more work in this area is needed to establish this approach as valid and clinically useful.

[1] Videogame addiction, and the larger category of media addiction, are highly contested constructs at the time of writing with some scholars, clinicians, and parents certain of its threat to child development and others certain that no such addiction can exist. We make no comment as to the reality of media addiction specifically in this work but note that scientific evidence has largely debunked many claims of risks of media use that fall short of addiction in recent years (Odgers and Jensen, 2020; Stiglic and Viner, 2019).

Technological-based therapy approaches, including telehealth approaches discussed in the previous section, are promising approaches to supporting behavioral therapy for individuals with ADHD. While the telehealth approaches discussed previously had a modest level of evidence to date, other approaches discussed in this section are still in the very early stages of testing in clinical populations. As the current literature contains preliminary studies, more research is needed to create prototypes robust enough to be used by multiple users at multiple times, and prototypes must be both cost-effective and durable. Randomized controlled trials should be conducted with these prototypes, using either a placebo control or an alternative behavioral therapy for comparison. Thus, without this level of evidence, most research in this area can be described as still emerging. Much more work is needed for these approaches to become a standard part of clinical care.

6.5   BIOFEEDBACK AND NEUROFEEDBACK

A widely studied strategy involves the practice of biofeedback (Basmajian, 1979; Schwartz and Andrasik, 2017). Biofeedback approaches involve providing biological information to patients in real time that would otherwise be unknown. This information can sometimes be referred to as augmented or extrinsic feedback, that is the feedback that provides the user with additional information, above and beyond the information that is naturally available to them, as opposed to the sensory (or intrinsic) feedback that provides self-generated information to the user from various intrinsic sensory receptors (Giggins, Persson, and Caulfield, 2013). For individuals with ADHD, there have been two main approaches to support self-regulation with biofeedback: controlling breathing and controlling brain activity, also known as neurofeedback.

Biofeedback, along with breathing exercises, can be used to slow down respiratory rates and promote relaxation (Giggins, Persson, and Caulfield, 2013). For example, ChillFish (Sonne and Jensen, 2016d) is a videogame controlled by a tangible thermistor sensor of exhalation. The game aims to control a pufferfish (the players' character) with a sensor, with the aim of collecting as many starfish as possible. When the player breathes, the fish inflates and moves. A pilot study with 12 children with ADHD showed that the breathing sensor was not appropriate for children to blow, and not all the children were able to use it, suggesting that more work is needed to understand how this approach might be beneficial and for whom it might be beneficial. The research team then went on to redesign ChillFish with a more robust sensor, but they had not yet evaluated this redesigned tool with children with ADHD at the time of this publication (Sonne et al., 2017).

Another example is DEEP (Van Rooij et al., 2016), a VR biofeedback game that allows players to explore an underwater fantasy world by using their breathing to control their movement (Steiner et al., 2014a, 2014b). Players wear a belt with a stretch sensor to measure breathing to help players relax. An ABAB design study was conducted with eight children from 12–16 years old with ADHD. During this time, participants completed five or six DEEP sessions. Results demonstrated a reduction in state-anxiety, but improvements in classrooms behaviors were only noted for half of the participants. Given the size and preliminary nature of the study, additional investigation with a larger sample size seems warranted and necessary.

Figure 6.2: One of the study authors, Mads Jensen, playing the Chill-Fish game from the Chillfish commercial site, http://www.chillfish.dk/ where the authors include instructions for building your own among other useful resources. Image courtesy of Mads Jensen.

On the other hand, neurofeedback training to support the self-regulation of brain activity for individuals with ADHD has been widely used and is considered a non-invasive approach to reducing ADHD symptoms (Arns et al., 2015; Steffert and Steffert, 2010; Marzbani, Marateb, and Mansourian, 2016; Rossiter and La Vaque, 1995; Linden, Habib, and Radokevic, 1996; Maurizio et al., 2014). It has been the subject of several meta-reviews (e.g., Sonuga-Barke et al., 2013; Cortese et al., 2016; Loo and Makeig, 2012; Lofthouse et al., 2012; Arns et al., 2009; Micoulaud-Franchi et al., 2014; Bussalb et al., 2019), and we addressed this work earlier in the book in a chapter on cognition (Chapter 4).

Learning how to control biological signals is challenging, but there is some evidence that technological interventions may help improve our ability to control our behaviors and minds. Although there is much research on neurofeedback, other types of biofeedback have not been as widely explored. Given that current technology can track physiological signals more easily than technologies available in the prior decade (e.g., smartwatches can sense heart rate and provide breathing training), there are ample opportunities to develop and study new intervention approaches. Indeed, although not the subject of our book, biofeedback has been considered for a variety of other health concerns, such as asthma (Lehrer et al., 2004), constipation (Heymen et al., 2003), headaches (Bunzynski et al., 1973), and psychiatric conditions (Glueck and Stroebel, 1975).

6.6   CONCLUSIONS AND FUTURE DIRECTIONS

Emerging technologies have enormous and largely untapped potential for the improvement of interventions to support behavioral management and self-regulation. New, more sophisticated tracking of physiological signals may allow researchers to infer and evaluate behaviors, which in turn can help promote self-regulation. Researchers should envision more efficient and robust methods for data collection in order to train models that support both the detection and evaluation of behaviors.

Additionally, prompting and behavioral interventions should be co-designed with clinicians, caregivers, and individuals with ADHD. In particular, the absence of research regarding technological support for adults with ADHD in this area is notable and concerning. Much self-regulation and behavior management develops in childhood. However, adults still struggle with impulsivity, hyperactivity, and inattention. The fact that many adults living with ADHD are often undiagnosed suggests that they have developed their own coping mechanisms, possibly with the support of mainstream technologies, such as meditation apps, mobile organizers or calendars, and mobile mental health support. On the other hand, it is possible that they continue to struggle, exert tremendous effort, and do not have the coping mechanisms or supports needed to help them realize their full potential. These areas should be a priority in future research, with an emphasis by researchers on what we can learn from existing strategies, including strategies identified or developed by adults with ADHD, and how we can augment or broaden the impact of these strategies for even greater gains for adults with ADHD.

Finally, we must continue to explore novel technological approaches. Virtual environments and robotic assistants may support behavioral therapies for individuals with ADHD in an engaging manner, particularly as they become more common and less expensive. More work is needed to explore how to co-design interventions using virtual reality environments so that distractions can be controlled to improve ecological validity. In summary, there is an untapped potential for technologies to support individuals with ADHD by helping them to recognize and evaluate their behaviors—so they can reflect on them and make decisions accordingly. We hope that this potential, as illustrated throughout this chapter, will inspire the current and future generations of computer and clinical scientists to develop and bring into practice evidence-based tools to enhance the lives of individuals with ADHD.

CHAPTER 7

Academic and Organizational Skills and Support

Success in schooling and the workplace are often highly dependent upon high levels of organization and self-regulation. Indeed, it is common for advertisements for jobs to emphasize organizational skills and for workshops preparing students for postsecondary education to emphasize such skills, whether for vocational school, two- or four-year colleges, or even graduate education. The sheer volume of self-help books available on organization (e.g., Paul, 2003; Selk, Bartow, and Rudy, 2015; Mackenzie and Nickerson, 2009) indicates the importance of organization for success and the challenges faced by many people. However, students and workers with ADHD may require additional support (Pinsky, 2012; Knight, 2019), including the use of ATs (Voytecki et al., 2009) to meet the expectations of instructors and employers. In Chapter 8, we focus on organizational skills as they relate to daily life and employment. In this chapter, we focus specifically on support for the academic environment.

7.1 ACADEMIC SUPPORT FOR STUDENTS WITH ADHD

Children with ADHD often need support in the classroom to address difficulties with executive functions, particularly attention, working memory, organization, and self-regulation. These difficulties can contribute to challenges in learning (including difficulty learning to read), organization (such as misplacing homework), working memory (including trouble remembering and following multiple-part directions), and classroom behavior (such as difficulty sitting still and paying attention). In the U.S., children with ADHD typically receive 80% or more of their instruction in a general education setting; in these settings, individualized intervention, support, and accommodations legally mandated by the U.S. Individuals with Disabilities in Education Act (IDEA) can often be delivered sporadically (e.g., Harlacher, Roberts, and Merrell, 2006; Nowacek and Mamlin, 2007). In other nations, this support can be more mixed (Agrawal et al., 2019), prompting the United Nations Educational, Scientific, and Cultural Organization (UNESCO) to issue a report in 2017 that children are less likely to complete primary or secondary school in many nations around the world if they have disabilities (UNESCO, 2017). This report further suggests that many systems fall short of living up to the UN Convention on the Rights of Persons with Disabilities (2006). Countries vary in how they approach this concern, some creating guidelines or laws nationally while others rely on efforts at a more local level. For example, in Canada, each Canadian province has its own policies to ensure free public education rather than national law; in the UK, a specific law nationally governs special education and identification processes; and in China, a law governs special education, but services are not always provided and diagnostics are mixed (Agrawal et al., 2019).

Regardless of the nation in which services and education are provided, perceptions of individualized support or intervention as impractical or time-consuming can be barriers to implementation (e.g., Harlacher et al., 2006). Additional barriers include variability in teachers' understanding of ADHD-related challenges and insufficient training in how to implement effective interventions to address difficulties (e.g., Pavri, 2004). These and other instructional barriers could be addressed in part using technologies designed to implement evidence-based school intervention strategies.

While most of the interventions reviewed in this chapter describe support for individuals with ADHD, class-wide interventions also have been explored and are summarized by Harlacher, Roberts, and Merrell (2006). They noted that supports can be provided for behavioral issues, such as contingency management and self-monitoring, as well as for more classically academic concerns, such as peer tutoring, instructional modification, and computer-assisted instruction. Unfortunately, research indicates that teachers in primary school may not provide sufficient accommodations for students with ADHD, and when they do, they are implemented in nonsystematic ways (Nowacek and Mamlin, 2007). Technology can be an equalizer in this regard, as computing systems tend to impose rigor and systemization in most contexts. The challenge then will be to ensure that valuable personalization and customization are not lost in the push for greater consistency. Likewise, such expanded access to accommodations, particularly those with technological elements, must come with additional training for both teacher preparation programs (Pavri, 2004) and continuing education.

Both practitioners and researchers have argued in support of the use of technology as part of an overall intervention strategy for ADHD (e.g., Murphy, 2005). Additionally, the use of AT has been shown to positively predict postsecondary educational attainment (Glynn, 2015). However, implementation in schools and universities has been mixed. For example, in a study involving 102 Swedish school children with ADHD and 940 without ADHD, Bolic et al. (2013) noted that fewer than half of the students with ADHD had access to a computer in the classroom and concluded that school staff should focus on enabling students with ADHD to use computers in their educational activities at school. In a U.S.-based study of data collected in the early to mid-2000s (Bouck et al., 2012), students with high-incidence disabilities like ADHD who received AT in school had more positive postschool outcomes in terms of a paid job, wages, and participation in postsecondary education. However, because AT was not a statistically significant predictor of positive postschool outcomes, the exact relationship, and recommendations for use, are still unclear. Notably, in this study, only 7.8% of participants reported receiving AT in high school, and only 1.1% after high school.

Anecdotally, these rates appear to be increasing and certainly more youths are now using smartphones and other technological tools in daily life. However, no comprehensive study has been conducted globally, and even if it had, the numbers would likely change so rapidly in a post-COVID-19 era as to make any study out of date nearly as soon as it was published. So, we are left to consider not how quickly computation can be taken up in schools, but how best to do so. Ofiesh et al. (2002) recommend use of AT devices for a broad range of postsecondary students with learning disabilities, including ADHD. Among these, they note that a strengths-based approach to AT might give greater access to education, that the context of a specific course and classroom should be considered, and that additional training is required as well periodic reviews with AT users.

Importantly, the idea that media use could precipitate ADHD-like symptoms has garnered a worrisome amount of press and popular attention (Ra et al., 2018; e.g., McRae, 2018; Howard, 2018). These reports indicate poor attentional control in high media use contexts, leading to lower reported well-being and diminished educational outcomes (Kushlev, Proulx, and Dunn, 2016; Rosen et al., 2013). However, more recent studies have failed to replicate the finding that ADHD and media multitasking are positively correlated, finding that the relationship did not approach significance (Fisher, 2016). Interestingly, engagement with computerized tools has also been recommended to treat "video game addiction" (Ruiz-Manrique et al., 2014).

In the remainder of this chapter, we highlight some of the ways in which technological tools have been developed to support the educational needs of students with ADHD. The majority of studies that have focused on academic skills for people with ADHD are centered on the experiences of primary school children. However, some recent studies have examined the relationship between technology and academic skills for adolescents and young adults pursuing postsecondary education. We discuss the current trends in developing and testing these tools and suggest directions for future work.

7.2   TECHNOLOGY SUPPORT FOR ACADEMIC SKILLS

Technological supports for academic intervention in children with ADHD include several techniques developed to improve reading skills. In one study of 21 students aged 10–12 years, visual coloring schemes improved reading through significant effects on attention span (Asiry et al., 2018). In another study, 20 postsecondary students with primary diagnosis of ADHD used assistive reading software for most of a semester (Hecker et al., 2002). Specifically, the software provided a synchronized presentation of text with auditory cues and incorporated highlighting and note taking features. The software improved reading speed by supporting the students in better attending to their reading, reducing their distractibility, and enabling them to read for longer periods of time. Despite these improvements, the use of the software had no effect on comprehension.

Technological interventions have also targeted the improvement of working memory, which is critical to learning. A computerized training program called CogMed RoboMemo is perhaps the most widespread tool used to try to improve working memory in children. One study randomized 60 youth (ages 12–17 years) with ADHD or learning disability (LD) into CogMed or another computer-supported condition. Adolescents in the working memory training group showed greater improvements than those in the other group, but there were no near or far measure effects (Gray et al., 2012). In other words, although they improved on working memory tasks, there was no evidence of improvement in academic settings. Nearly all of the 57 Swedish schoolchildren in another study saw positive effects of working memory training, which in turn appeared to be beneficial to their reading comprehension skills (Dahlin, 2011). In another study of RoboMemo with 67 children with ADHD, improvements were shown in multiple areas, with those in reading still in place 8 months later (Egeland et al., 2013). Finally, in a study of CogMed (Chacko et al., 2014), which was tested with 85 children with ADHD, researchers observed improvements in working memory storage but no differences for other outcome measures. Given these somewhat inconsistent findings and the large body of research on computerized working memory interventions, several meta-analyses and reviews have been conducted to assess the state of the field, and these are discussed in Chapter 4, which is focused on cognitive training.

More sophisticated and emergent tools are now being used to support literacy in children with ADHD. For example, Luna et al. (2018) developed a prototype of an augmented reality system to support literacy and vocabulary attainment in school-aged children. They pilot-tested this prototype with 25 users, finding the system generally usable and promising for future work. As another example, robots engaged in collaborative learning with children appear to have the potential to improve engagement with reading over long periods of time (Jimenez et al., 2016). Reflex is a mobile training application that uses computer vision to support interaction with physical devices (Spitale et al., 2019), building on the popular OSMO commercial platform (https://www.playosmo.com/en/). With this approach, the applications can use the built-in camera on a mobile device coupled with a mirror to point down at a table that holds physical objects. The application perceives these objects and renders them digitally on the screen. These tools, while preliminary, show some promise for academic learning. It is unclear what the mechanism is that underlies these successes. However, the potential for engaging other sensory stimuli, such as tangible in the case of Reflex, as well as the attention of a social agent—whether robotic and physical or digital and virtual—are likely factors that would support learning. Using these tools, children who may struggle to learn in one modality can use another while remaining engaged despite the potential for distractibility.

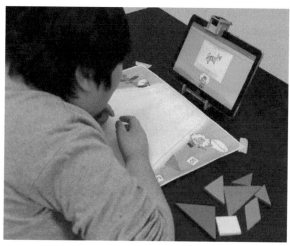

Figure 7.1: A participant doing the Tangram activity on the Reflex platform. Image courtesy of Mirko Gelsomini.

Despite these promising developments, there is limited evidence for the long-term efficacy of such approaches. For example, Project CLASS (Rabiner et al., 2010), which reported results from a randomized-controlled trial to study the impact of computer-assisted instruction (CAI) on attention and academic performance in 77 American first graders (typically 6 or 7 years old), there were demonstrated improvements through CAI in reading fluency and in teacher ratings of academic performance. In this study, the authors used a commercially available tool, Captain's Log from Braintrain® (https://www.braintrain.com/), that provides structured activities to support children training a variety of skills related to attention. Specifically, the authors chose to use ten exercises that train the ability to sustain auditory and visual attention. An example they provide in the publication requires children to press the space bar each time a symbol appears that matches another symbol on-screen. However, the effects of this intervention were absent by second grade. In this case, the control students also declined in attention problems by second grade, thereby erasing the differences but leaving open the possibility that long-term effects may still be seen in future years or with additional intervention.

Although improvement in reading is a well-studied area, there are also some efforts focused on supporting mathematics education. In earlier research, researchers (DuPaul and Eckert, 1998; DuPaul and Stoner, 2003) demonstrated that classroom-wide content could be broken into smaller chunks to improve math performance and off-task behavior (Ota and DuPaul, 2002). As another more recent example, Mautone et al. (2005) found that schoolchildren ages 10–12 years could improve math skills through the use of CAI. Five years later, Lewis et al. (2010) found that accessible digital math textbooks were preferable to print for both teachers and students. Additionally, the use of these textbooks tended to correlate with higher test scores. As in reading skills, augmented reality has shown some promise in supporting quantitative literacy. For example, Tobar-Munoz et al. (2014) designed an inclusive augmented reality game supporting Logical Math Skills Learning. They tested the game with a set of 20 students with diverse learning needs, including ADHD, suggesting that an augmented reality game can support children with special needs in learning such skills. However, this research is still in the early stages, and larger randomized trials are needed to clarify the benefits for individuals with ADHD.

It can be difficult to disambiguate technologies that support academic skills more broadly. Certainly, in the primary, secondary, and postsecondary markets, technology has an increasingly outsized presence. This has been especially true during the COVID-19 pandemic, as many schools have shifted toward delivering part or all of their education virtually, using a wide variety of technological tools. In this section, we focused only on those studies specific to academic support for students with ADHD. However, students with ADHD almost certainly experience educational technologies differently than those without ADHD; thus, in the section focused on online learning, we also review those studies available that articulate some of these differences.

7.3   TRANSITIONING TO POSTSECONDARY INSTITUTIONS

Simple off the shelf technologies as well as those that are more advanced and innovative can support adolescents and young adults as they transition to postsecondary education (Hayes et al., 2013). In particular, transitioning to postsecondary education can mean a loss of both structure and accommodations that students with ADHD have come to rely upon. Thus, a variety of experts recommend using AT and other technological solutions to support these students (Connor, 2012; Webb et al., 2008).

Given the challenges associated with organization, self-regulation, and time management for students with ADHD, it is perhaps unsurprising that both commercial products and research projects have examined solutions to supporting greater organization for college students. In one such study, three students with learning disabilities (LD) and/or ADHD were studied in a case study design; results indicated that students using a handheld computer, which we would now call a mobile device, reported increased reliance on the device compared to the student using a traditional paper planner, and that the digital approach was more proactive in supporting the student than the reactive paper-based approach (Shrieber and Seifert, 2009). In more recent work, the authors of this book (Cibrian et al., 2020a; Doan et al., 2020) found that smartwatches also could play this role, with children and adolescents appreciating the organization and prompting that calendaring and schedules in a smartwatch platform can provide. One might ask whether these tools could in turn create dependence on such devices. However, students with ADHD should be treated in this regard no differently than other busy students and professionals who regularly depend on planners, digital calendars, and other supports to remember assignments and schedules.

In their review of the literature for postsecondary transition, Mull and Sitlington (2003) suggested that the transition processes must include a focus on funding. While funding for people with disabilities has not necessarily improved in the intervening 17 years, technology prices have reduced dramatically and technological solutions have become more mainstream and accepted. Other recommendations remain largely true today, such as their recommendations that the selection of the specific ATs should address both student needs and environmental context and should providing training opportunities for students to learn to use the technologies.

The transition process must also include a focus on assessment of the learning technology needs of students prior to their attendance. In a survey of 142 students at a highly competitive university, Parker and Banerjee (2007) found significant differences between the technology needs, preferences, and fluency of undergraduates with and without disabilities. In particular, while all students reported high levels of comfort and fluency with technologies overall, students with disabilities indicated lower exposure to online and blended courses than their non-disabled peers. These findings indicate that early assessment of the needs might enable postsecondary institutions to provide access to and training with online learning environments prior to enrollment. The additional opportunities and challenges of such online environments are discussed in more detail in the following subsection.

7.4   ONLINE LEARNING ENVIRONMENTS

For the last 30 years, students with learning disabilities have been using AT more and more to access a wide variety of services, including those provided by other mainstream or learning technologies (Bryant and Bryant, 1999). During the 2020 COVID-19 pandemic, the use of these online learning environments and technologically enabled learning tools expanded massively and rapidly all around the world. Although we anticipate an eventual return to in-person instruction for most people, it is likely that increased reliance on such technologies, tools, and platforms is here to stay. Understanding how students with ADHD perform in such environments, the unique challenges they may face, and the opportunities these platforms afford will be essential to delivering equitable education in the future. In particular, the COVID-19 era has demonstrated that reforms requested by learners over the years focused on the challenges of traditional classrooms (e.g., Fovet, 2007) are now possible to address with distance learning technologies and other related approaches.

In an effort to address the particular challenges for college students with ADHD with regard to developing proficiency with technological tools used in postsecondary schools, such as learning management systems, and the self-regulation required while processing information online, researchers developed and piloted the Learning Technologies Management System (LiTMS). This approach uses five components to deliver services.

- **Universal Design for Instruction** (Scott et al., 2001; Rose et al., 2006), which advocates a design approach in which instructional technologies and instructional materials are designed to be usable by all people, to the greatest extent possible.

- **Authentic Learning Need**, which intends to encourage students to think deeply, considering difficult questions and proactively raising them, to consider multiple forms of evidence and recognize the nuances in them, and to navigate difficult problems.

- **Preferences**, which can include the ways in which students prefer to receive information as well as behavioral aspects like study habits.

- **Range of Learning Technologies**, which includes consumer products, personal, and ATs.

- **Individualized Technology Design**, which suits the unique needs and wants of the people using the technologies.

Although this model has not been tested robustly, it is well-grounded theoretically and matches the best practices for user experience design (Parker et al., 2009b).

Despite many schools, colleges, and universities making use of approaches like Universal Design, challenges remain. For example, in a study of 14 first-year undergraduate students across the Georgia state university system, all with learning disabilities and some with ADHD (exact number unspecified by the authors), Wimberly et al. (2004) identified technical barriers (such as computer malfunctions), design barriers (such as usability and readability), and intrapersonal barriers unique to individual students as all preventing them from high-quality experiences with education-based information technology. At the same time, even some well-designed strategies showed no effect in practice. For example, Cho (2004) tested seven strategies for promoting student self-regulated learning as described by Zimmerman and Martinez-Pons (1986). Although Zimmerman, Bonner, and Kovach (1996) argued that students' self-regulation can be taught and improved through student effort, Cho's technology-enabled interventions showed no effect.

Online education can be particularly difficult for people with attention disorders. The content is delivered through an inherently distracting device, and few, if any, social and physical constraints exist to encourage attention. Indeed, the authors observed in their own practices that following the rapid switch to digital education due to COVID-19 quarantines or school facility closures, children and adolescents with ADHD struggling to succeed in a virtual learning environment. VibRein attempts to address these challenges through a mobile application that uses sensors to provide "continuous supervision" and encourage attending to the video through haptic feedback (Toshniwal et al., 2015). Surveillance through the camera tracking the student is required of an approach like this. However, facial recognition in schools has become a hot ethical and practical question in recent years (Andrejevic and Selwyn, 2020; LoSardo, 2020). Similarly, as students and instructors are coping with the widespread use of this kind of visual surveillance for remote learning during COVID, the challenges to ethical and equitable use of such technologies are becoming more evident (Doffman, 2020; Halaweh, 2020). Increasingly, both students and instructors are revolting against the use of cameras in this way in educational environments (e.g., Hubler, 2020; Kelley, 2020; Swauger, 2020; Nicandro et al., 2020). This trend indicates that researchers and product designers will need to develop new, less invasive approaches to provide the same kind of engagement or abandon these types of prompting technologies altogether.

Limited effects and reported challenges may be due to the mismatch in some cases between the ways in which most designers create learning technologies and the optimal learning context for people with ADHD. In a case study of an individual student with diagnoses of both autism and ADHD, three types of disorientation were identified—navigational, contextual, and procedural—each of which required unique strategies for its mitigation (Meyers and Bagnall, 2015). Navigational disorientation refers to difficulty navigating what the authors call "hypermedia," but which we now identify as simply the way the modern web works. Interconnected links that may take us down long paths that do not result in the answers we seek have long been a challenge of good information architects. While Google's original mission of "organizing the world's information" has increasingly become one of helping people to search the inherently disorganized information of the world, this kind of information structure can be exceedingly disorienting to learners with ADHD. Brown argued in 2009 that students with ADHD might become "lost in hyperspace" due to the inherent challenges of this media approach (Brown, 2009). Contextual disorientation involves difficultly placing the information found in this complex web within a larger context. Learning management systems at universities as well as the instructor themselves, attempt to do this work, but it can still be challenging for students. Finally, procedural disorientation refers to the difficulty that students may have in understanding precisely what task they are meant to complete next. Increasingly, learning management systems have attempted to address this challenge for all learners by allowing instructors to group materials into "modules" or "units," by making greater use of calendaring across courses and other interventions. However, instructors often use these features inconsistently, which can create an even more problematic learning environment for a student with ADHD who needs consistent structure to be successful in an online learning environment.

At the same time, there is a strong indication that online learning environments have the potential to improve the context of learning for some students with ADHD. Graves et al. (2011) interviewed 11 students with ADHD and/or learning disabilities enrolled in STEM courses. Students reported that using asynchronous online access enhanced their learning experiences according to six themes: clarity, organization, asynchronous access, convenience, achievement, and disability coping mechanism. This has been supported in the anecdotal experiences of students with ADHD who transitioned to distance learning during COVID-19. While many have reported challenges, as noted above, many of these individuals also report advantages. For some, the convenience of learning at home was beneficial, as it allowed for more frequent breaks, more variety in physical learning environments (such as being able to stand, walk, or change positions while "in class"), and self-paced learning.

To date, research efforts indicate a variety of potential disjunctions between the characteristics and properties of the learner's online learning environment and the specific learning needs and preferences of each individual learner. Thus, inclusive online learning environments in higher education must take into account this need for personalization and adaptability. In a comprehensive report from 2012, researchers at the Ontario College of Art and Design University set out to identify opportunities to use innovative technologies to support postsecondary education for students with ADHD. This report focused on software but was informed by interviews and observations as well as a comprehensive review of the literature. In their cataloging, they found a wide variety of time management software—such as those we describe in Chapter 8—but a scarcity of mood- and motivation-focused software. They also identified an emerging set of software focused on integrating with existing learning tools, which led them to conclude that "leveraging the features of stronger software categories will strengthen weaker software categories, especially when these features can be correlated with adult ADHD needs." They went on to describe a set of tools they view as particularly needed, including speech APIs for search, time visualization software, and smart goal management tools. In the intervening eight years, some of these areas have certainly developed more strongly, but the opportunities are currently unrealized in others.

7.5   CONCLUSIONS AND FUTURE DIRECTIONS

Our review of the current literature indicated that the majority of technological tools that aim to support academic skills were designed to be used by children, adolescents, and adults with ADHD or their parents. Far fewer tools are available to support instructional staff, such as classroom teachers and university professors. While some of these tools may be useful for students with ADHD in the context of both classroom and online learning, much less attention has been paid to the notion of holistic learning environments and support for multiple stakeholders.

There exist dozens of technological tools and mobile applications available to teachers that aim to support general student success and classroom management, which may also benefit students with ADHD, but the literature supporting the effectiveness of these tools for students with ADHD is scant. Certain mobile applications for schools have been touted in the press as "promising practices" (*The Zones of Regulation*®; Katz, 2012). One tool, *Class Dojo* (https://www.classdojo.com/, accessed February 17, 2020), which developers claim is implemented in 95% of schools in the U.S., includes a mobile application designed for teachers to manage classroom behavior and also aims to improve communication between parents, teachers, and students. Much work has yet to be done to determine if these technological supports are in fact, helpful when working with students with ADHD and whether or not they contribute to meaningful outcomes. As communication between home and the school remains a paramount challenge in managing symptoms of ADHD, research should specifically address how these tools affect the quality and frequency of teacher and parent/caregiver communication.

Additionally, education, both in the primary and secondary school as well as higher education, is an industry in which adaptation and inclusion of tools are often done quickly and in response to particular contexts at the level of an individual course or department. Thus, institutions and researchers alike have limited insight into the efficacy of such tools and approaches at scale. Ofiesh et al. (2002) argue that service-providers must collect and share data regarding effectiveness. This approach would certainly enable much more large-scale naturalistic data collection and identification of trends in these data, though randomized controlled trials would provide greater potential for identifying causal relationships. The current shift to online educational approaches that occurred in response to COVID-19 presents endless opportunities to study at a widespread level what went well and what did not. Collecting and sharing data across countries and the world could propel our understanding of how to best employ technology in educational settings.

CHAPTER 8

Life and Vocational Skills

Adults with ADHD experience a variety of challenges when it comes to life skills, employment, and other transition and post-transition needs, and their needs can be radically different from those of children who tend to be more often the focus of research in this space (Ginsberg et al., 2014). In particular, research has indicated that adults with ADHD are less likely to succeed in the workplace than those without ADHD (Coetzer, 2016). The Multimodal Treatment Study of Children with ADHD (MTA) followed children with ADHD (original N = 579 with 82% retention into adulthood) and a local normative comparison group (original N = 258 with 93% retention into adulthood). Early adulthood evaluations were conducted 16 years after enrollment in the study, and participants were, on average, close to 25 years old. In early adulthood, there were significant differences between individuals with ADHD and controls in impulsivity, mood (rapid exaggerated changes in mood), educational attainment, occupational outcomes, and sexual behavior outcomes (Hechtman et al., 2016). More specifically, there were significant differences in whether or not an individual had obtained postsecondary education, the times he or she had quit or been fired from a job, current income, and whether or not he or she was receiving public assistance (Hechtman et al., 2016). This research points to the critical importance of developing and studying tools to support positive outcomes in adulthood. In this chapter, we first provide background on the experiences of adults living with ADHD, then review the technological systems that have been studied or suggested for supporting people with ADHD. We begin that review with general life skills and then move into the broad employment and vocational skills domain.

8.1 LIVING WITH ADHD

The concerning results from the long-term MTA follow-up studies point to the long-term negative outcomes associated with ADHD, but also provide some hope for individuals with ADHD, their families, and intervention or clinical service professionals. Among the participants diagnosed with ADHD in childhood, by the time they reached adulthood, 50% (n = 226) had persistent ADHD symptoms, but for 50% (n = 227), symptoms had desisted, no longer reaching the threshold for ADHD. Those with persistent symptoms experienced greater challenges overall, but for those whose symptoms had desisted, outcomes fell somewhere in-between the symptom-persistent group and the local normative comparison group (individuals without ADHD who had been matched on other variables) (Hechtman et al., 2016). This result highlights the importance of continuing to develop tools to support individuals with ADHD while adjusting the context around them to better accommodate their experiences in hopes of contributing to better long-term outcomes.

Although incredibly important to the success and long-term quality of life, few studies have addressed life skills and technology engagement explicitly. Those research projects that do exist address a range of domains, although pre-dominantly address issues of self-regulation and daily management. Daily activities, in particular, can be a challenge for people who struggle with disorganization and impulsivity. Fujiwara et al. (2017) proposed an Internet of Things (IoT) solution to help support the daily activities for adults with ADHD, focusing on helping them remember activities and find lost objects. Although others have focused on computational systems for finding lost objects (e.g., Peters et al., 2004; Nakada, Kanai, and Kunifuli, 2005), this project was focused on supporting people with ADHD and involved an evaluation with professionals who work with people with ADHD that showed some promise. Specifically, the authors presented the application to four professionals who work in the field and followed up the demo with a questionnaire. The results indicated that the professionals had some belief that the proposed solution would work and be helpful, but how well would depend on the individual. These approaches can be useful for initial evaluations of tools to reduce an unnecessary burden on a clinical population that is already overburdened and can be difficult to recruit. However, such evaluations should not be considered sufficient in lieu of direct engagement with people with ADHD before a tool is deployed or produced on a larger scale.

In another example of using promising approaches from other domains to support daily activities, Bul et al. (2015) developed a serious game called "Plan-It Commander" to promote behavioral learning, time management, planning, and organization for children with ADHD. They tested the game using a survey of 42 participants, finding general satisfaction and indicating a need for greater research (Bul et al., 2018). Robotics are recently emerging as a promising approach to the challenge of limited human resources for behavioral interventions; however, results are still preliminary (Clark-Turner and Begum, 2017).

Pagers, a precursor to many of the mobile and wearable devices we now use, showed early promise. For example, a fourth-grade boy with ADHD found success through custom paging to support his daily routines (Epstein et al., 2000). Technology has advanced dramatically since the time of pagers, and now smartwatches are a particularly promising platform for the delivery of daily support in people with ADHD. Dibia (2016) developed Foqus, a smartwatch platform for encouraging mindfulness and self-regulation in adults with ADHD. Cibrian et al. (2020a) found that children with ADHD are interested in having support through smartwatch technology, and in a related project, Doan et al. (2020) described a prototype tool called CoolCraig that translated components of an existing evidence-based program (e.g., Lakes et al., 2009, 2011) into a smartwatch format. This work is ongoing and focused on harnessing the potential of smartwatches in a way that supports self-regulation and daily activities in individuals with ADHD.

Figure 8.1: Screenshots of Foqus demonstrating three different modes that the application can display, including Pomodoro, Meditate, or general Health Tips. Image from ResearchGate, provided by Dibia Victor.

Cognitive Assistive Technologies (CAT), such as those described by LoPresti, Mihailidis, and Kirsch (2004) have shown particular promise in supporting people with a variety of life skills, not simply related to ADHD. CAT (Mihaillidis and LoPresti, 2006) uses devices and tools to re-habilitate and augment user's cognitive disabilities or to translate a challenge into a set of activities and techniques that can better match the user's strengths. Lindstedt and Umb-Carlsson (2013) studied an intervention using 74 CATs over 15 months with adults with ADHD. Users valued weekly schedules, watches, and weighted blankets the most. The authors note that CAT is most valuable when individually tailored and structured to support the specific needs of the individuals, a set of conditions most easily met by a combination of technologies that allow for personalization and the involvement of trained professionals.

Patterns in AT use are difficult to discern and likely vary widely by geography, socioeconomic status, and disability status (Sonne et al., 2016b). Using survey data from 163 postsecondary disabil-ity service providers, the most commonly used assistive technologies in 2001 were voice recognition systems, reading machines, frequency modulation systems, and text enlargement (Ofiesh et al., 2002), with many of those systems most commonly being used by those with vision and hearing impairments. Such surveys must be undertaken frequently with the ability to break down usage precisely by a variety of categories, such as those listed above.

Beyond the executive function and self-management approaches developed to support people with ADHD, some researchers have explored technologies for highly specific needs. In some cases, parents are already using technologies in novel ways, such as a family who used loca-tion-based services to monitor their child, who had dual diagnoses of ADHD and Asperger's Syn-drome (Thomas, Briggs, and Little, 2010). In most cases, however, the literature reports on specific applications developed by or in conjunction with researchers. In one promising study, Bruce et al.

(2017) described a computer application (Drive Smart) to improve hazard detection in drivers with ADHD. Preliminary results with 25 individuals with ADHD were promising, with improvements in hazard detection noted after a single training session and maintained over a 6-week period. Similarly, Clancey, Rucklidge, and Owen (2006) found that virtual reality may be useful to teach adolescents about road-crossing safety, particularly given that in their study of 48 children aged 13–17 years (24 with ADHD), those with ADHD walked slower, showed lower margins of safety, and in general demonstrated behaviors that put them at higher risk for collisions.

Moreover, medication adherence, long a target for technological support, can be improved using a mHealth app, such as the one described by Weisman et al. (2018) and tested with 39 children with ADHD. In this study, the authors found higher overall pill counts (a common measure for medication adherence) in patients who used the app. Family daily routines are also an area of ongoing struggle in families affected by ADHD. Sonne et al. (2016a, 2016c) developed a tool, MOBERO, to support families living with at least one child with a diagnosis of ADHD during the children's morning and bedtime routines. They then conducted a follow-up study one month later (Sonne et al., 2016a) after removing the AT from the homes of 13 children with ADHD. Their study used technology to reduce parent frustration and conflict levels around morning and bedtime routines, leading to improvements that lasted after the removal of the technology. These studies provide examples of how novel technologies can address specific needs and support daily functioning and life skills in individuals with ADHD and their families.

8.2   CAREER READINESS AND JOB SEEKING

A precursor to successful employment and engagement in the workplace for anyone is career readiness. Certainly, basic training on technologies, to be used in both job searches in general (e.g., job search and resume websites) or those for a particular job (e.g., the Microsoft Office suite), should be taught anyone with and without barriers to employment. However, in this section, we particularly focus on those studies of technologies to support this kind of career readiness and job seeking for people with ADHD.

In a one-year ethnographic study of job seekers with cognitive impairments, including ADHD, in Norway, Michelsen, Slettebo, and Moser (2017) found that mandating the integration of technology into career activities as well as the use of the students' own technologies supported the introduction of Cognitive Support Technologies (CST) for adults with ADHD and/or Asperger's. In this work, the researchers included a broad range of tools in their definition of CST, such as tools used professionally in companies for productivity and communication as well as those technologies that might help job seekers to improve their competitiveness or become more motivated to cope with other life challenges, including smartphones and a variety of apps installed on those phones for organization and communication. This particular project built on other research that shows CST for employees and students with cognitive disabilities, including autism, ADHD, and other developmental disabilities, can be effective (e.g., Gentry et al., 2012; Hill et al., 2013); however, the authors did not disambiguate which technologies were effective in which ways for which students. In follow-up work, these authors (Michelsen, Slettebo, and Moser, 2019a, 2019b) found that the introduction of ATs to the rehabilitation support process for adults with ADHD was valuable both in terms of individual empowerment of participants and the long-term completion of such programs.

Furthermore, in a promising examination of embedding technology instruction into existing curricula, Lombardi et al. (2017) added IT literacy modules into an existing online transition curriculum for secondary education in the U.S. for students with disabilities, including ADHD. In a quasi-experimental study across six secondary institutions, high schools in American parlance, the intervention group improved IT literacy, and the control group made no such gains. These results show promise both for integrating such instruction into existing transition programs and for teaching IT literacy using an online platform that by definition, requires some limited IT literacy before beginning.

8.3   EMPLOYMENT AND WORKPLACE SUPPORT

Several reviews of the challenges of people with ADHD and employment already exist. Schaeffer, Thomas, and Hersen (2004) described the literature alongside their own clinical experience in a book focused on both learning disabilities and attention deficits in the workplace. Harris (2020) described the documented challenges of adults with ADHD in the workplace in her doctoral dissertation before delving into the findings of a small qualitative study. Shahin et al. (2020) reviewed the literature on workplace participation for young adults (ages 18–35 years) with brain-based disabilities, including ADHD, finding that environmental domains strongly influenced workplace participation, with the majority of the studies focused on systems and organizations and other studies referencing social support, physical access, and availability of assistive technology. In a more practical guide, Honeybourne (2019) presented a mix of research literature and practical advice aimed at helping employers build and maintain a neurodiversity-friendly workplace.

The search for employment itself relies on a level of organization and networking that is difficult for anyone, regardless of an ADHD diagnosis. However, those with ADHD often report feeling isolated, with challenges in interaction and communication with job coaches, career counselors, therapists, and others (Adamou et al., 2013; Pitts, Mangle, and Asherson, 2015). Notably, in a survey of 210 adults, 105 with ADHD, Pitts et al. (2015) found that respondents with ADHD tended to become disengaged from the process and disaffected by the employment system overall, in part due to the challenges of matching a job with their needs, abilities, and interests. Finally, job searches increasingly are conducted online, as evidenced by the authors' experience working on a transition to work task force in education and clinical practice. However, in many cases, the support structures that we have created in educational and clinical environments have not evolved as fast as we might like with this change. Thus, there is an opportunity to develop approaches to using technologies effectively in the job search as also discussed in the previous section on career readiness.

Even if an appropriate position is found and an interview is secured, people with ADHD often continue to struggle. Some studies have found that adults with ADHD are less accurate in describing their past employment history than peers, resulting at times in an exaggeration of certain skills or experiences (Cartwright, 2015; Steinau and Kandemir, 2013). As an added challenge, adults with ADHD must also determine whether to disclose their condition at the time of interviews, which could improve chances if the employer understands the condition and its treatment, but is more likely to introduce a negative bias into the interview process (Adamou et al., 2013). Young adults in one study even reported that requiring fewer physical accommodations seemed to increase their success rate in acquiring a job (Toldrá and Santosb, 2013). As in employment searches, there is an opportunity here to use technology to support the interview process, and some research projects have begun to do this for populations other than ADHD (Hayes et al., 2015; Ulgado et al., 2013; Ogbonnaya-Ogburu et al., 2018).

Once employed, there are opportunities to support people with ADHD as the challenges associated with employment become sometimes more complex (Chang and Edwards, 2015; Verheul et al., 2016). In a sample of 448 employees from across Puerto Rico, Rosario-Hernandez et al. (2020) found that ADHD had a direct effect on both task performance and counterproductive work behaviors. However, they saw no relationship between ADHD and organizational citizenship behaviors. Well-designed and structured employment with appropriate support from supervisors and managers combined with cognitive-behavioral and pharmaceutical treatments can mitigate the challenges associated with ADHD symptoms in the workplace (Barkley, 2013; Biederman et al., 2005; Chang and Edwards, 2015; Martinez-Raga et al., 2013; Seli et al., 2014). Adults with ADHD often struggle with procrastination (Fletcher, 2013; Bozionelos and Bozionolis 2013), time management, organization, and prioritization (Schafer et al. 2013). Similar observations have been noted among young adults (ages 17–24 years old) as well as in the broader adult population (Nguyen et al., 2013; Young et al., 2017). At the same time, a lack of understanding by those without ADHD in the workplace can make it difficult for those with ADHD to work with their colleagues in teams (Barkley, 2013; Fuermaier et al., 2013; Klein, Mannuzza, and Olazagasti, 2012; Lopez et al., 2013; Surman et al., 2013; Morris et al., 2015) or to have the type of workplace friendships that many people enjoy (Almasi, 2016). These challenges present opportunities for technologies to support those without ADHD in understanding those with the disorder as well as to help those with ADHD increase their success in the workplace, as detailed in this chapter. In particular, people with ADHD may demonstrate enhanced creativity and the ability to deal effectively with complexity (Bozionelos and Bozionelos, 2013; Schnieders, Gerber, and Goldberg, 2015) as well as resilience (Reid, et al., 2014), all of which can be leveraged with additional supports, such as inter-active technologies, to improve workplace interactions for people with ADHD.

8.4   MENTORING AND COACHING

A key aspect to support for transition to employment for young adults with ADHD is coaching. Thus, the notion of job coaches or employment support is embedded in the majority of transition programs. Employment-focused coaching tends to be highly individualized, focusing on goal-set-ting, organization, planning, and time management and workplace skills (Kubik, 2010; Prevatt et al., 2011). Gerber, Ginsberg, and Reiff (1992) studied 71 successful adults with learning disabili-ties, finding that coaches, mentors, and supportive coworkers were key to their eventual success in the workplace.

Prevatt and Yelland (2013) studied a variety of aspects of the coaching process for people with ADHD, including but not limited to coaching processes and structures and how they correlate to success. In this work, ADHD coaching was conducted over a five year period with 148 college students. Clients who received the 8-week intervention showed significant improvement in study-ing and learning strategies as well as in their self-esteem and satisfaction with school work. Tuttle, Ahmann, and Wright (2016) found that coaching supported beneficial client outcomes, particularly when used in conjunction with CBT and/or medication. Tucha (2017) also found coaching to be one of the nonpharmacological interventions that can be successful in the treatment and support of people with ADHD, particularly as they transition into adulthood.

However, human coaching simply does not scale to the number and diversity of people who can benefit from this kind of engagement. Thus, people have begun to examine options for technologically enhanced supports that produce similar outcomes to human coaching. In one such study, Burke et al. (2013) used VideoTote, a tablet-based video modeling and prompting tool to support young adults in the workplace with autism, ADHD, and related conditions. In this study, participants and parents both gave the software high ratings, and pre and post-tests indicated that the prompting and modeling techniques were useful supports. Such studies should be expanded and tested with people with ADHD without autism to understand the full potential for individ-uals with ADHD. The VideoTote application was available for a time in the App Store for free. However, it has since been removed, demonstrating the challenge with this kind of research. There are many potential reasons for the loss of such a commercial product, including but not limited to determining the market demand for such technologies. General execution should a commercial approach be undertaken includes all of the challenges of launching a business (e.g., management of the product lifecycle, marketing, sales, human resources, and finance). Finally, another challenge may be the lack of desire by researchers to undertake a commercial or even non-profit product enterprise. Steady streams of research revenue can sometimes help research teams to pick up the challenge of maintaining tools shown to be effective. However, even this approach requires an investment on the part of the research team that may not be possible as the ability to publish the work wanes over time, students and post-doctoral scholars move on, and so on. These challenges may be best addressed by encouraging more collaboration between researchers and business communities with interest in picking up and maintaining and updating products over time.

Figure 8.2: VideoTote user interface, showing the user selecting no stops, which allows for watching the video from start to finish. The other option, "stops," supports stopping the video automatically at each step or chapter. Image from ResearchGate, uploaded by Keith D. Allen.

8.5   CONCLUSIONS AND FUTURE DIRECTIONS

There is substantial evidence that people with ADHD can be successful in the workplace when employers create "ADHD friendly" environments that allow their employees to use the behavioral strategies they have developed to help them succeed (Fields, Johnson, and Hassig, 2017). Such accommodations might include working from home, which has been found to be effective for people with traumatic brain injuries (Roessler et al., 2017) as well as ADHD and other disabilities (Solstad Vedeler and Schreuer, 2011). Of course, working from home requires the ability to successfully use telework technologies, which many employers have declared impossible in the past. The COVID-19 era has demonstrated that such accommodations are possible for a wide range of employees and should be supported whenever possible; recent media outlets have included reports from executives and researchers suggesting that expanded opportunities for telework are here to stay and will fundamentally change the "workplace" for decades to come. This potential outcome could benefit many individuals, including those with ADHD, who may especially benefit from the flexibility to create a work environment conducive to their needs. Other technologically based workplace accommodations might include the use of dictaphones, dual monitors, assistive devices for communication, and computerized phones and alarms. Such tools have positively impacted work satisfaction and work maintenance (Ripat and Woodgate, 2017; Morash-Macneil, Johnson, and Ryan, 2018; Solstad Vedeler and Schreuer, 2011), and researchers have shown that a variety of off the shelf technologies can be used to support adolescents with barriers to employment, including ADHD (Hayes and Hosaflook, 2014; Hayes et al., 2013). In 2008, Lazarus (2008) described how we might better incorporate people with disabilities in virtual workplaces. Such a study should be undertaken again post-COVID-19 to understand how people with disabilities fared in the workplace during this difficult time.

Time management and self-regulation are particular challenges that repeatedly emerge in the literature as well as in self-reports from people with ADHD, making support essential to creating an ADHD-friendly work environment. Nadeau (2015) found that adults with ADHD who used environmental modifications that limited internal and external distractions (e.g., a job that is physically active or integrates exercise into the daily routine) benefit substantially in terms of their ability to self-regulate and reduce feelings of restlessness and hyperactivity. Dipeolu, Hargrave, and Storlie (2015) similarly found that people with ADHD could manage their symptoms using strict schedules and minimizing distractions in their immediate work environments. These accommodations are all relatively straightforward to provide, particularly for office work. However, Morgensterns et al. (2015) found that employers and coworkers rarely considered how crucially important such accommodations are to employees with ADHD.

Technological tools for adolescents, young adults, and adults should be a high priority, especially as they are likely to need support in the transition into adulthood, making this a prime period for the implementation of self-management and vocational success strategies. None of the studies we reviewed that had clinical trial evidence addressed needs associated with the transition from adolescence to adulthood. In spite of this, there are promising interventions in development and early testing. Although some tools used highly innovative technological solutions, such as robotics and VR, the vast majority of research in this space has been conducted on technologies that could realistically scale with the off the shelf products currently available. Additionally, this space, more so than the literature in any other chapter in this book, focuses on adults who are more able to participate in large-scale, longitudinal research trials. Given the disparate adult vocational and life outcomes documented for adults with ADHD (e.g., Hechtman et al., 2016), is it is imperative that researchers prioritize the large-scale clinical trials necessary to demonstrate efficacy of life, vocational, and post-transition supports. Such trials would make governmental funding in many nations easier to obtain for these supports as well as insurance funding, for example, in the U.S.

CHAPTER 9

Motor Control, Physical Accessibility, and Physical Activity

Although the majority of research surrounding ADHD focuses on issues of cognition, attention, and working memory, the physical experience of ADHD for both children and adults is well worth considering. It may be a motor or physical behavior that first suggests ADHD, an issue discussed in more detail in Chapter 3 (Computationally Supported Diagnosis and Assessment). Likewise, physical activity is increasingly considered as an intervention to address ADHD symptoms, in earlier years as a strategy to reduce hyperactivity and more recently as a strategy to improve attention and other executive functions. In this chapter, we describe the background in assessing and understanding motor control and ADHD and introduce some research on physical activity as it relates to ADHD symptoms. We then move into the specific issues of technology related to motor behaviors and physical accessibility, first by describing technologies that promote motor skill development and motor control, specifically, and then describing work that promotes physical activity.

9.1 MOTOR CONTROL AND ADHD

Motor and cognitive development are fundamentally intertwined (Diamond, 2000). Therefore, it is not surprising that difficulties in cognitive development are associated with difficulties in motor development. Children with ADHD diagnoses often have different and slower trajectories related to their motor control than their typically developing peers (Gilbert et al., 2011). They may not meet developmental milestones involving timed repetitive and sequential movements, unintentional movements accompanying voluntary tasks, and balance. Fine motor skills are often delayed, and skills like producing neat handwriting can be especially difficult. Even much earlier research in ADHD acknowledged differences in motor development, indicating that children with ADHD tended to develop accurate rhythm later than their typically developing peers and showed immature patterns of motor overflow later (Denckla and Rudel, 1978).

A variety of tests are available for evaluating motor control in all children and have been used to assess children with ADHD. These include tests focused on motor control and movement (Denckla, 1985) as well as those focused on manual dexterity and balance (Henderson and Sudgen, 1992). This battery assesses the manual dexterity, ball skills, and balance of children. Some tools are more typically used in schools and clinical settings (Deitz et al., 2007; Mancini et al., 2020) while others require clinical intervention, such as Transcranial Magnetic Stimulation (TMS) (Gilbert et al., 2011).

Challenges with motor control in people living with ADHD can present in a variety of ways and have varied impacts on life experiences. The exact mechanisms by which motor challenges manifest in some children with ADHD are unclear. Such challenges are inconsistent across the population, leading to high variability in both life experiences and treatment. A variety of explanations for the motor skills differences in children with ADHD and their peers without ADHD can be given, including but not limited to comorbidity, inattention, or lack of inhibition (Kaiser et al., 2015). Comorbid developmental coordination disorders may explain motor coordination difficulties for some individuals. One study found that the motor skills of children with only ADHD —and no other diagnoses—did not differ from a control group without ADHD. Indeed, in this study, the presence of reading disabilities were better predictors of ADHD than motor impairment of some kind (Kooistra et al., 2005). Likewise, a study including 84 children—20 with ADHD only, 42 with ADHD and another co-occurring disability or illness, and 40 children with multiple co-occurrences and ADHD—found that the presence of co-existing diagnoses had a significant influence on both cognition and motor behavior in children with ADHD (Crawford et al., 2006).

A variety of treatments for motor control and physical accessibility challenges have been studied and used for children and adults with ADHD for many years. These include both behavioral and pharmaceutical treatments, but given the emphasis of this book on technological support and behavioral augmentation through technologies, we here do not address pharmaceutical approaches.

9.2   TECHNOLOGICAL APPROACHES TO SUPPORTING MOTOR SKILL DEVELOPMENT AND CONTROL

In addition to efforts to increase engagement and participation in physical activity, which may in turn promote motor development and control, there is research describing a range of technological tools designed to specifically support motor skills and motor control development. This work includes computerized cognitive training, robotics, and interfaces to support physical interventions. We describe some of this work in subsequent subsections.

9.2.1   COMPUTERIZED COGNITIVE TRAINING TO SUPPORT MOTOR SKILLS

Since the dawn of the personal computer, computers have been used to support a variety of diagnostic and therapeutic goals. However, they are less frequently used specifically for motor skills, given their limitations in input and output. Typing, using a mouse, and translation from a three-dimensional world into a two-dimensional space are all difficult cognitive and motor skills, particularly for children.

Computers can, however, be used to support growth and skill development in the very areas in which their use can be challenging. For example, Eliasson et al. (2004) demonstrated that children with ADHD move the mouse more slowly, perform more jerky movements, and produce more errors than children without ADHD. This type of finding can help develop models to detect symptoms of ADHD that might otherwise be missed and design corrective assistive technologies that can, for example, improve the performance of children with ADHD when mousing through, smoothing the interaction experience dynamically in the background. Similarly, in a study testing handdrawn movement, 62 children used an electronic pen on a digitizing tablet to join targets. The results of this study indicated that children with ADHD showed no temporal difficulties in the tasks but did demonstrate subtle spatial difficulties (Johnson et al., 2010). In particular, they showed a subtle spatial bias toward the right that the children without ADHD did not show, despite all children both with and without ADHD being right-handed. This kind of subtle bias is an indicator of the potential for accessibility features in computing systems that might support stylus and mouse use with adaptive interactions that adjust for such subtle biases.

One of the longest available technologies to support motor control in people with ADHD and others with related conditions is The Interactive Metronome® (IM). This training and assessment tool was developed to improve cognition, attention, and focus, as well as motor and sensory skills. The interactive metronome was developed in the early 1990s and included a "Main Station" that connects to a computer and measures rhythm, auditory feedback delivered by headphones, and one or more devices that the client uses to play the game—such as buttons to press or a mat to tap with a foot. This digital program is used in homes, schools, and clinics for adult rehabilitation (e.g., as related to brain injuries and neurological diseases) as well as pediatric practices focused on ADHD, autism, sensory and learning disabilities, and speech and language delays. Clinicians use IM to play digital interactive games alongside functional therapy interventions to explicitly target the neural timing within and between the brain regions. The tool uses a game-like platform to instruct and engage the user while providing instant feedback and logging progress. The IM training has been used successfully with boys with ADHD (Shaffer et al., 2001) to improve attention and academic skills, but also for this chapter most notably motor control. In smaller studies, such as one from Korea involving two boys with ADHD (Namgung et al., 2015), the IM was found to facilitate timing, attention, and motor control in the boys who used it. In a similar study by Park and Choi (2017), these results were replicated two years later with two other boys. A variety of other studies have shown similar promise (Bartscherer et al., 2005; Gu et al., 2017; Park and Kim, 2018). However, to date, no comprehensive large-scale study of the IM for children or adults with ADHD has been reported.

9.2.2   PHYSICAL INTERACTIONS IN MOTOR SKILL SUPPORT AND ASSESSMENT

Teacher shortages, particularly for special education, have become a hallmark of the modern educational system. A wide variety of important skills, particularly in early childhood, require hand-over-hand or other close one-on-one contact. However, very few children have the attention of a teacher or an aid for the entire day. Furthermore, with the COVID-19 global pandemic remaking schooling as we know it, this kind of close contact is even more difficult. Robotics offers a promising alternative to these challenges. A social and physical assistant in the form of a robot can provide one-on-one attention to augment that of a teacher or parent. For example, Palsbo and Hodd-Szivek (2012) found that a 3D robot-assisted repetitive motion training system used for 10 hours of training improved handwriting fluidity for children with ADHD as well as speed for those referred for slow writing speed.

Similar to a robotic approach, other tangible interactions can provide some of the benefits of hands-on instruction without the need for another teacher or aide in the classroom. In a series of studies, Shih et al. (2011) tested this approach using a gyrostatic mouse and a Nintendo Wii Remote Control to detect hyperactivity and spontaneous standing (Shih et al., 2011, 2014; Shih, 2011). In these cases, the researchers automatically detected the behavior and then used a gyroscopic mouse or a Wii controller—devices that allow a user to control a piece of software by moving the device through the air—to provide or remove vibrotactile feedback—used in this case as an aversive—when the nonpreferred behavior was observed. In a small study, this approach appeared promising for supporting individuals with ADHD in modifying their behavior to be considered more socially acceptable for the context. Although the technologies used here may be effective in teaching the student with ADHD to change their behaviors, researchers should also consider developing technologies for such environments (e.g., classrooms) that help all students and teachers to feel more comfortable when the student with ADHD needs to self-regulate through nonpreferred behaviors.

Beyond this kind of simple vibrotactile feedback, other researchers have explored notions of other types of interaction more broadly. For example, Polipo, a tangible toy created for children with neurodevelopmental disorders, enables a highly customizable experience with children and therapists able to personalize its light, sound, and tactile stimuli (Tam, Gelsomini, and Garzotto, 2017). Other open-ended tangible environments have been shown to support children with disabilities in learning, though the evidence remains limited (e.g., Bakker et al., 2011; Garzotto and Gonella, 2011).

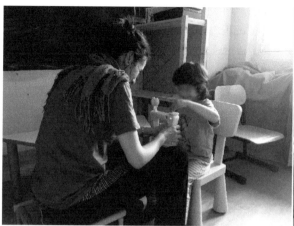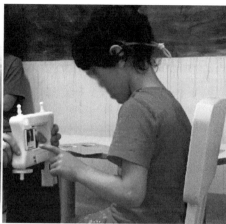

Figure 9.1: A child plays with Polipo, on the left seated with a therapist, and on the right a view of the child and Polipo alone. Images courtesy of Mirko Gelsomini.

Gesture-based interfaces—in which the software or device is controlled using specific gestures, like scrolling, pinching or tapping—provide much of the same appeal as tactile systems, allowing groups to customize and balance their interactions beyond what a traditional keyboard, mouse, or even touchscreen might do. For example, Sharma et al. (2018) conducted a series of studies using three gesture-based applications to support motor coordination (HOPE), joint attention (Balloons), and life skills (Kirana). Although not empirically validated yet, their work is particularly compelling as it considers a context under-represented in the research literature, living with ADHD in India.

Finally, the last frontier in off-the-desktop interaction may be virtual and augmented reality. These technologies can be combined with physical interventions to support improvement in motor function. For example, Shema-Shiratzky et al. (2019) conducted a study with 14 non-medicated children with ADHD during 18 training sessions over 6 weeks, in which the children walked on a treadmill while avoiding virtual obstacles. Their results indicated that parents perceived children's behavior to improve meaningfully and that gait regularity significantly increased during dual-task walking. Executive functions improved following training, but attention did not. These results, while preliminary given the pilot nature of the study, suggest that further work in this area is warranted.

9.3   PHYSICAL ACTIVITY AND ADHD

Physical activity has long been considered important for motor development and physical health, but in recent years, there has been growing body of evidence suggesting that physical activity can be beneficial for cognition (e.g., Vazou et al., 2019; Diamond, 2015), which introduces the possibility of physical activity producing multiple benefits—improved physical fitness, motor coordination, and cognition—particularly for individuals with ADHD who may struggle in one or more of these areas. For example, Verret et al. (2012) found that children with ADHD in a ten-week exercise training program improved their muscular capacities and motor skills as well as performance on information processing tests. Berwid and Halperin (2012) argued that there is evidence supporting the hypothesis that physical activity impacts structural brain growth and functional neurocognitive development, and this could have profound implications for intervention to alter the trajectory of ADHD. They noted at the time that most physical activity research with children with ADHD had been pilot research, with limitations including small sample sizes and unblind status of research- ers or raters. In spite of these limitations, there appears to be growing evidence to suggest that a variety of physical activity interventions could directly improve executive functions in children with ADHD, but more research is needed before interventions become a recommended treatment for ADHD. This interest physical activity interventions for children with ADHD, is particularly relevant given that the risk of overweight and obesity is also associated with ADHD (e.g., Waring and Lapane, 2008), and positive physical activity habits might maintain or improve physical health while also reducing ADHD symptoms and promoting the development of motor skills.

As the benefits of physical activity for physical health and motor skill development are generally widely known, here we focus on providing some background for the more recent work on the relationship between physical activity and cognition, particularly as it may be relevant to the development of technological tools to promote physical activity. As the relationship between physical activity and the brain has received increasing attention, models have been proposed to ex- plain the neurobiology of benefits to cognition. Best (2010) stated, "There are at least three general pathways by which aerobic exercise may facilitate executive functions in children: (1) the cognitive demands inherent in the structure of goal-directed and engaging exercise; (2) the cognitive en- gagement required to execute complex motor movements; and (3) the physiological changes in the brain induced by aerobic exercise." However, the evidence indicates that not all forms of exercise benefit executive functions equally (Diamond and Lee, 2011; Vazou et al., 2019; Diamond, 2015), and that "the degree to which the exercise requires complex, controlled, and adaptive cognition and movement may determine its impact on EF [executive functions]" (Best, 2010). Recent research (e.g., Vazou et al., 2019; Pesce et al., 2020) has demonstrated that physical activity programs that involve greater cognitive engagement (involving cognitive challenges or mindful components) pro- duce stronger effects on executive functions than less cognitively engaging aerobic activities (such as simply jogging). Diamond (2015) argued that moving without thought produces little sustained change in executive functions and that practices that require both thought and movement—are likely to have a stronger positive effect on executive functions and deserve further study.

It is worth briefly considering interventions that integrate physically active and mindful components to target symptoms of ADHD, particularly as this may inform the development of technologies that could support interventions with multiple components. Because mindfulness practice involves efforts to self-regulate attention, as awareness of this practice has grown, there has naturally been an increasing interest in the potential of mindfulness training to improve executive functions (particularly attention) in individuals with ADHD. Mitchell, Zylowska, and Kollins (2015) concluded that while preliminary studies of mindfulness training for children and adolescents with ADHD have produced promising results, the research to date has had methodological limitations, including small sample sizes and lack of active control groups. Intervention combining low-intensity physical activity with mindfulness training have been described in the literature; for example, Zylowska et al. (2009) described how to train attention during sitting or walking meditation by focusing attention on an anchor (such as breath), observing that distractions occur, and letting go of them, and refocusing on the anchor. This work is interesting as Vazou et al. (2019) demonstrated that in typically developing children, physical activities with a mindful component (such as yoga or martial arts) produced stronger effects on cognition than less cognitively-engaging forms of physical activity. However, to date, little research in the areas of physical activity and mindfulness interventions has been conducted with individuals with ADHD.

In addition to studying physical activity as an intervention to promote physical and cognitive outcomes, research has also examined the impact of movement on learning and classroom performance. Results of a study conducted by Hartanto et al. (2015) showed that in children with ADHD, performance on an executive function task was better when children were active during the test, supporting the hypothesis that physical movement may function as a compensatory strategy to increase arousal and attention in children with ADHD. Many years earlier, Etscheidt and Ayllon (1987) reported that even five-minute bouts of exercise could reduce hyperactive behaviors during classwork. Thus, a commonly recommended classroom accommodation for children with ADHD has been allowing "physical activity breaks" to reduce restlessness and increase on-task behaviors. In spite of these promising findings, there is no current data demonstrating how much or what type of physical activity is required to have a meaningful impact on ADHD symptoms, but interest in this line of research appears to have grown in the last decade.

9.4   TECHNOLOGICAL INTERVENTIONS TO PROMOTE PHYSICAL ACTIVITY

Given the substantial interest in physical activity for individuals with ADHD, it is not surprising that technological interventions are being developed and tested to assess potential benefits on ADHD symptoms. As noted above, physical activity is associated with improved cognitive functioning in a wide range of people and has been investigated specifically to support people with ADHD. Technologies can broaden the potential for physical interventions by making such activity more fun and engaging, allowing for home-based intervention, and producing positive outcomes while tracking progress.

For example, Weerdmeester et al. (2016) demonstrated that a full-body-driven videogame could reduce teacher-reported ADHD symptoms in 73 children with ADHD as well as influence positively both impulsivity and gross motor skills. Taken further, these types of whole-body interactive games are often referenced collectively as "exergames," and they provide a promising approach for treatment and support of individuals with ADHD. Exergames have particular promise in potentially being able to target cognitive and physical processes at the same time. Benzing et al. (2018) demonstrated improvement in reaction times in inhibition and switching with just 15 minutes of exergaming, in which children performed moderate intensity physical activity. In this case, the exergame used was a commercial game called "Shape Up" from Ubisoft on a Microsoft XBOX Kinect platform.

Benzing and Schmidt (2019) described and studied a cognitively and physically demanding exergaming intervention. The exergaming intervention was cognitively engaging (e.g., requiring attention, inhibition, switching, and speed of action) as well as physically challenging. In a well-designed randomized controlled trial with 51 children (ages 8–12 years), Benzing and Schmidt (2019) compared 8 weeks of home-based exergaming intervention to a waitlist control and found significant improvements in executive functions (e.g., reaction time for inhibition and switching), parent ratings of psychological difficulty (i.e., Conner's 3 Global Index Scores), and motor skills, although significant improvements in ADHD symptoms were not observed. Their research uniquely highlights the potential benefits of integrating cognitively engaging and physically challenging exergames into daily routines for children and adolescents with ADHD that is likely to inspire further research given a widespread focus on the potential of exercise to benefit executive functions in general and for ADHD, in particular. However, they recommended the need for customized exergames for this population in order to better target ADHD symptoms in particular (Benzing and Schmidt, 2019).

Some researchers have begun to examine customizing exergames, in particular, in support of children with autism, another neurodevelopmental disorder. For example, FroggyBobby, is an exergame designed to support improved motor control and attention (Caro et al., 2017). In this work, the authors focused on using a participatory design approach to support this custom design. Although they did not test FroggyBobby with a large sample size, the teachers involved in this work perceived that the exergame had some impact on motor functioning as well as other behaviors, such as socialization and body awareness. Similarly, Caro et al. (2016) proposed a full room, technologically instrumented, to support exergaming: the ExerCaveRoom. This approach might provide children with a wide variety of developmental disabilities the opportunity to practice customized therapeutic interactions while being recorded for assessment toward progress.

Beyond these custom systems, a vast array of technologies and research studies have focused on exergaming for children with autism (e.g., Caro et al., 2017; Peña et al., 2020; Cibrian et al., 2017, 2020b). Although these technologies were not, in most cases, tested with children with

ADHD as well, it is worth quickly reviewing them for what they can provide us in thinking about exergaming more broadly for children with a variety of disabilities, but particularly those that are neurodevelopmental in nature. Certainly, this book, with its focus on ADHD, does not attempt to fully review such approaches for other populations, such as individuals with autism. Notably, some studies, such as Hilton et al. (2014, 2015), indicate substantial promise in improving working memory, metacognition, and the motor skills of strength and agility for children with autism. Researchers should certainly take these positive results and not only replicate them with a broader group of children with autism but also expand this work to both children and adults with ADHD.

In an attempt to make some sense of the many preliminary studies, Fang et al. (2019b) produced a systematic review of exergaming for children with autism. In this review, they found that children with autism showed significant improvements in physical fitness, executive function, and self-perception, but there were minimal effects on emotional regulation and little to no effects on motor development. These findings are somewhat in contrast to the studies cited above that do show improvements in motor skills for children with ADHD. Whether these differences are due to the preliminary nature of all of these studies or whether to differences between these clinical groups has yet to be determined. Researchers should seek to both scale up these interventions and to examine differences amongst various groups that are, in other ways, potentially related.

Children with ADHD might particularly benefit from exergame interventions that enable them to play with other children with a variety of abilities, disabilities, and health conditions. In a systematic review of exergames for children and adolescents who are not typically developing, the authors found strong evidence that active video games improved balance, with more limited evidence related to coordination, running, and jumping (Page et al., 2017). In one study, children with cerebral palsy were more willing to engage in physical therapy, for example, when they were able to play exergames with other children, both with and without cerebral palsy (de Greef et al., 2013). Certainly, schools, that are inherently mixed ability venues, could benefit from these kinds of approaches in their adaptive physical education programs.

Moreover, applications already embedded in mobile technologies (e.g., smartwatches, smartphones) could be used to help promote physical activity in individuals with ADHD. In our own work with youth with ADHD, we are exploring how to promote physical activity with smartwatches as part of a comprehensive parent-child system designed to support healthy behaviors, such as physical activity and sleep, while also providing ADHD-specific behavioral intervention for both parents and children (Cibrian et al. 2020a).

9.5   CONCLUSIONS AND FUTURE DIRECTIONS

Although ADHD is most commonly associated with executive function difficulties, physical challenges, including difficulties with motor coordination, can be present and often have a negative impact on day-to-day functioning. Motor control is, in fact, such a prevalent issue for children with ADHD in particular that a battery of tests has been used to study motor control in children with ADHD. The exact mechanisms for these challenges are unclear, but what is clear is the potential for physical activity to simultaneously address motor and cognitive functioning. Thus, this chapter outlined the growing rationale and evidence for technological interventions to promote physical activity and simultaneously target motor and cognitive functioning.

This chapter points to several critical directions for future research. Exergames and other approaches that influence and increase physical activity in children and adults with ADHD have shown promising results in preliminary studies, and there is a growing scientific rationale for paying more attention to how physical activity, motor functioning, and cognitive functioning are inter-twined. In response to this promise, a wide range of technologies—from exergaming to robotics to mobile applications—are being developed and studied. Likewise, other tools have been developed to specifically focus on specific motor skills, including both simple desktop or web applications, as well as augmented reality and wearables. In a period in which a worldwide pandemic and social distancing have reduced opportunities for both physical activity and to some degree therapies focused on motor functioning, the potential for affordable, home-based technology to fill the gap is especially important.

Although the work is promising, large-scale randomized controlled trials are very few, as is often the case in the early stages of development. The range of technologies is inspiring, and their application to critical challenges in ADHD is promising. We hope that now and in the future, greater financial investment will yield the types of intervention studies needed to eventually bring these technologies into the marketplace where they can directly benefit individuals with ADHD. Perhaps in this current time of social distancing, studies could be undertaken using existing tools (such as exergaming) both to promote opportunities for home-based physical activity and to propel the field forward through widespread efforts to better understand the potential of existing technology.

CHAPTER 10

Discussion and Conclusion

ADHD is one of the most common neurodevelopmental disorders in childhood and often persists into adulthood. It is commonly undiagnosed in adults and broadly considered as heritable (Khan and Faraone, 2006), making the experience of ADHD in families particularly challenging. While people living with ADHD are often successful, they may still struggle with a wide variety of symptoms and challenges. Additionally, as an invisible disability, it can be hard to seek and receive the accommodations necessary for this success. Technological supports, particularly those that can be built into everyday life or allow for personalized engagement at scale are particularly appealing given that context and those, along with the wide variety of diagnostic and assessment tools, were overviewed in this book.

10.1 MULTIDISCIPLINARY RESEARCH AGENDA

All of the authors of this book have been working in multidisciplinary research around the health and wellness of children and adults with neurodevelopmental disorders for many years. Two have been collaborating from these different worlds for a decade. The authors collectively have written numerous literature reviews, including three on sub-topics covered in this book. Despite the years of trust, engagement, and communication practices built in that time, reviewing this literature was still an incredibly difficult task.

Multidisciplinary research in any field struggles with issues of how to conduct research, where to publish, and what constitutes a contribution to the field. In the case of ADHD, the basic science required to move the field forward includes the development and dissemination of clinically relevant science, construction of robust theoretical and empirical models, and the design and development of novel technologies. Likewise, true progress comes when evidence-based decisions drawn on this research are the norm, and products in the marketplace have been rigorously tested for efficacy. Translating the scholarship of this field into real-world practice requires the tech transfer inherent to getting computer science research into the development pipeline as well as the translation of behavioral and clinical research into policies, procedures, and practices.

Some simple but tractable challenges in this kind of scholarship include: publication venues, keywords, style of research, and authorship. Clinical research must be indexed by PubMed to be accessible to practitioners, medical and psychology students, and to be included in meta-reviews. However, despite some progress—largely driven by leaders in the American Medical Informatics Association, the Association for Computing Machinery, and the IEEE—many computing journals are still not indexed in this important reference site. Furthermore, even those that are can be nearly impossible to find and understand, as computing articles tend not to use the standard abstract presentation that many clinical venues do, making these articles functionally inaccessible, if not physically inaccessible to clinicians and clinical researchers. On the other hand, clinical journals, although they include randomized studies, often lack the technical descriptions needed to replicate technological tools. Moreover, technological solutions are often designed from a clinical perspective, leaving aside the needs of the end-users, in this case, individuals with ADHD, something the human computer interaction (HCI) field is trying to improve. This is only an example for U.S.-based research. These issues are augmented when the multidisciplinary team research also includes researchers from all over the world (not only from the "Global North"), as each country tends to have its own policies on which publication venues are acceptable by their standards.

While more open tools like Google Scholar have improved this access, keywords still remain problematic. Simply put, computing researchers like to change keywords a lot with the new technology trends and nearly never use the terms "technology" or "computer" that are too broad to be useful in computing specific fields. Without these keywords, however, ADHD researchers and practitioners are unlikely to stumble onto an article that invokes technical jargon for keywords like "deep learning," "Internet of Things," or "wearables." And yet, computing researchers need those in other fields to pick up the work as it becomes unpublishable in computing venues around the time it becomes eligible for a clinical trial and publication in a behavioral or medical context.

The notion of what is publishable brings us to the concern regarding publication style. This book overviews a wide variety of papers in both content and structure. With such an enormous set of publications available to read, people are always going to tend toward the ones that are structured in familiar ways. However, a radically multidisciplinary space such as this should bring researchers from computing, design, education, psychology, psychiatry, physiology, vocational rehabilitation, and more. We, as scholars, must commit to reading in these fields even when uncomfortable as well as learning to write in those styles. Finally, reviewers and editors must become more open to norm-violating publications with scholarly merit.

Most academic researchers are keenly aware of the authorship norms in their own communities. However, these are highly differentiated across fields. Nowhere does this become more apparent than when writing a major review in a field such as this. The authorship lists in the references in this book vary from a single author to more than 20. As the field grows and as greater numbers of technical contributions to work appear in the form of creation and maintenance of large-scale datasets, code repositories, and algorithms, researchers will likely find themselves needing to address authorship concerns in radically new ways. In our own collaborative work over the last decade, we have learned to map out publications ahead of time, have uncomfortable but frank conversations about authorship, and take any challenges that emerge with a generous and open spirit.

Although this book does some of the work of addressing these multidisciplinary challenges by going beyond a standard structured literature review, there is still work to do. Technological venues allow for—in fact, demand—work that is innovative and new, which clinical journals rarely allow, opting instead for a full randomized controlled trial (RCT) before publication. So, while replication studies, studies at scale, and any long-term study that might demonstrate true efficacy but has no technological innovation can be incredibly difficult to publish in a computing venue, anything shy of a fully functioning product—which requires major investment—can be nearly impossible to get in the hands of clinicians. These are not problems to be solved by single research teams, no matter how excellent their interdisciplinary approach. These are major structural and systemic issues that must be addressed for this field to truly advance.

10.2   INCLUSIVE AND ENGAGED RESEARCH

Due to the complexity and sheer number of people of all ages, gender identifications, and ethnicities living with ADHD, a vast array of technologies are regularly brought to bear in support of this community. At the same time, these technologies can be used to oppress, surveil, and dismiss people with ADHD. This mix of power for opportunity and growth and risk for abuse and toxicity is not unique to technologies in this space. However, the history of humankind's responses to people who think differently, live with mental health challenges, and cope with ADHD indicates that special care should be taken in considering how the technologies we develop might be studied, received, and used.

One of the ways to deal with the challenges of ensuring access, inclusion, and equity in our research is to include people with ADHD in our research teams and labs. The authors of this book all make a practice of inclusion of people with a wide variety of disabilities into our research programs. These individuals cannot be engaged simply as volunteers or community partners, though those are valuable roles. They must also be invited into paid positions, studentships, and full team memberships.

Beyond the good practices of individual research teams, given the power to both support and harm individuals with ADHD and other mental health conditions, technological systems should be subject to regulation. For the last several years, in the U.S., the Food and Drug Administration (FDA) has been examining whether and how to regulate digital health applications (Shuren et al., 2018). The outcome of this ultimate regulation will likely slow the development—particularly commercial development—of some digital products in this space. However, the overall effect of this slowdown will also likely be technologies with greater evidence of their efficacy before going to market. One can hope that changing the speed of development may also increase the ability for research labs and companies to create neurodiverse teams for the design, development, and testing of these products.

To meet the high bar of the FDA and other regulatory agencies, the community will have to develop higher quality standards for evaluation. Returning to the issues of interdisciplinarity from the prior section, these standards will have to cross communities for them to be effective. As of now, almost no standards reach across academic communities in this way. Even relatively standardized concerns that have regulating bodies, such as the ethical treatment of human subjects, are still not consistently managed. A true commitment to technological tools that are efficacious and ethical, however, will demand such standards emerge, not just in technologies for mental health but across the scope of computing. Moreover, researchers from all over the world should be acknowledged, and, as researchers (in both clinical and computer fields), we cannot assume that technology that is developed in one context may work for others that clearly have different resources and needs. So, although the FDA is a well know regulatory agency in the U.S., we must look beyond those norms, especially for developing countries.

Finally, computing researchers, especially those who work in the space of technologies for health, will have to develop a more consumer-facing approach to their scholarship. Open-source software communities offer one model for ensuring wide access to the products developed in government and university research labs. However, the use of such code repositories often still requires heavy technical skills. A world must be developed in which research is accessible to people with ADHD and those who care about them, whether through open access publishing, a notion of a "generic" software intervention like generic prescription medication, or other mechanisms.

10.3   HUMAN AUGMENTATION OR REPLACEMENT

The world is simultaneously experiencing a surge in computational power and a highly limited workforce in terms of people who can care for those with ADHD or other mental health challenges. In such a scenario, the temptation to replace humans with agent-based and other intelligent support is enormous. However, recent seminal books on psychotherapy research have illustrated the tremendous importance of a therapist-patient relationship as a common factor that predicts psychotherapy outcomes above and beyond improvements attributed to specific therapeutic models or strategies (Wampold and Imel, 2015). The importance of this human connection should not be discarded in the race to deliver technologies to expand access to mental health intervention. We see possibilities in a balanced approach in which computational elements support those things that technology is best at supporting, and therapists, parents, friends, and educators continuing to play key roles in supporting individuals with ADHD. An apt metaphor for this approach might be the monitoring systems in a hospital that do the drudgery of checking the patient's vital signs frequently, recording progress, and ensuring no crisis is occurring, while clinicians do the high-level work of monitoring these monitoring systems, responding if something changes that is problematic, listening to patients, and providing direct treatment and patient education when needed. Likewise, schools are already making use of intelligent tutor software programs to provide customized computer-based instruction overseen by a teacher. We noted in some papers published in technological scientific venues that programs were not described with this balance in mind. For future work, we suggest that this balanced approach is taken; in striking this balance, researchers and developers should acknowledge the strengths and limitations of technological approaches, which in turn may increase the willingness of clinicians to adopt and use technologies, with the understanding that the tools will support and extend their impact, rather than fully replace them. To achieve this balance, multidisciplinary teams are especially important, as they simultaneously will draw from expertise in clinical science/practice and computational science.

10.4   THE NEED FOR TRANSLATIONAL RESEARCH AND SYSTEMIC CHANGE

In this book, we have frequently noted the gap in translation from design to adoption among clinicians, families, and other systems (schools, workplaces). The computer science literature describes a wide variety of exciting, novel approaches to treatment and assessment. Most of these papers describe development and feasibility or user testing with small sample sizes. In the clinical sciences, the randomized clinical trial (often with a hundred or more patients) is considered the gold standard for evaluating an intervention. Yet, few of the tools developed by computer scientists have been studied at this level, as their goal is first to test prototypes to understand how the users use them, how they can be improved, and discard those prototypes that have no potential from the early stages, with the aim of only testing the efficacy of those prototypes with more possibilities to success. This inevitably limits the uptake of products among clinicians, families, and individuals with ADHD.

We noted that despite the innovative tools described in the literature, very few are commercially available at this time. What is needed to move these ideas into the world so that they can directly benefit individuals with ADHD and those around them? Certainly, we believe that multidisciplinary and diverse teams are a necessary first step. Other systemic changes in the way we conduct multidisciplinary science research are also necessary. For example, there is a need to increase the representation of these early-stage studies in clinical journals to increase awareness of the potential value of technological tools and to inspire multidisciplinary research agendas, especially among clinical scientists. This will require some adjustments and greater flexibility in journal editorial policies, which we believe are possible to achieve without sacrificing scientific rigor. Another greatly needed systemic change involves how we fund clinical trials once a prototype is available for this type of testing. Clinical trials of sufficient sample size are extremely expensive to conduct, often costing several million dollars over a few years. In the U.S., clinical scientists remain largely dependent on clinical trial funding from the National Institutes of Health, yet this funding is extremely limited, competition for funding is fierce, and it can take years to successfully obtain funding. Often, by the time funding is received, substantial modifications to the study design are needed as technologies have advanced in the interim. These modifications generally are a good thing, though, as they can help improve the study design and product studied. Changes could include more NIH support for digital health technology as well as bringing other investors and grant-making entities into this work.

We encourage investigators and the community at large to become engaged in advocating for systemic changes that will decrease this substantial lag in the time from development to the marketplace. Our current environment in the time of COVID-19 indicates that this is possible when the demand and motivation is there. For example, telehealth capacity has existed for decades in the "Global North," and there have been a number of studies published over the last decade or two that set the framework for how to conduct a telehealth practice ethically and safely. However, clinician uptake of telehealth practices was extremely limited prior to March of 2020. Since that time, many organizations and clinicians have rapidly engaged in telehealth practices. In our own experiences at the University of California, we saw how rapidly departments were able to transition to telehealth when the environment demanded it, and we are witnessing how effectively we are able to deliver mental health services and reduce barriers to treatment using technologies that have been available to us for at least the past decade. We hope that we will learn from this experience and that it will fundamentally change for the better the way that we approach mental health care and multidisciplinary work in the years to come, particularly as it relates to more rapid engagement with technologies that can improve clinical care.

Correction to: Research Advances in ADHD and Technology

Franceli L. Cibrian, Gillian R. Hayes and Kimberley D. Lakes

Correction to:

F. L. Cibrian, *Research Advances in ADHD and Technology*
https://doi.org/10.1007/978-3-031-01606-6

Author affiliation country is incorrect for Gillian R. Hayes and Kimberley D. Lakes. Affiliation country should be USA not UK for authors Gillian R. Hayes and Kimberley D. Lakes. This has now been corrected.

The updated online version of this book can be found at

https://doi.org/10.1007/978-3-031-01606-6

F. L. *Cibrian, Research Advances in ADHD and Technology*
© Springer Nature Switzerland AG 2023
https://doi.org/10.1007/978-3-031-01606-6_11

References

Aase, H. and Sagvolden, T. (2005). Moment-to-moment dynamics of ADHD behaviour. *Behavioral and Brain Functions*, 1(12), 1–14. DOI: 10.1186/1744-9081-1-12. 17, 58

Aase, H. and Sagvolden, T. (2006). Infrequent, but not frequent, reinforcers produce more variable responding and deficient sustained attention in young children with attention-deficit/hyperactivity disorder (ADHD). *Journal of Child Psychology and Psychiatry and Allied Disciplines*, 47(5), 457–471. DOI: 10.1111/j.1469-7610.2005.01468.x. 17, 58

Abibullaev, B. and An, J. (2012). Decision support algorithm for diagnosis of ADHD using electroencephalograms. *Journal of Medical Systems*, 36(4), 2675–2688. 33

Adamou, M., Arif, M., Asherson, P., Aw, T.-C., Bolea, B., Coghill, D., Guðjónsson, G., Halmøy, A., Hodgkins, P., Müller, U., Pitts, M., Trakoli, A., Williams, N., and Young, S. (2013). Occupational issues of adults with ADHD. *BMC Psychiatry*, 13(1), 59–74. DOI: 10.1186/1471-244X-13-59. 81, 82

Adams, R., Finn, P., Moes, E., Flannery, K., and Rizzo, A. (2009). Distractibility in attention/deficit/hyperactivity disorder (ADHD): The virtual reality classroom. *Child Neuropsychology*, 15(2), 120–135. DOI: 10.1080/09297040802169077. 24

Agrawal, J., Barrio, B. L., Kressler, B., Hsiao, Y. J., and Shankland, R. K. (2019). International policies, identification, and services for students with learning disabilities: An Exploration across 10 countries. *Learning Disabilities: A Contemporary Journal*, 17(1), 95–113. 65, 66

Alchalabi, A. E., Shirmohammadi, S., Eddin, A. N., and Elsharnouby, M. (2018). FOCUS: Detecting ADHD patients by an EEG-based serious game. *IEEE Transactions on Instrumentation and Measurement*, 67(7), 1512–1520. DOI: 10.1109/TIM.2018.2838158. 17, 27

Alchalcabi, A. E., Eddin, A. N., and Shirmohammadi, S. (2017). More attention, less deficit: Wearable EEG-based serious game for focus improvement. In 2017 IEEE 5th International Conference on Serious Games and Applications for Health, SeGAH 2017 (pp. 1–8). DOI: 10.1109/SeGAH.2017.7939288. 17, 27

Almasi, N. G. (2016). The comparison of self-efficacy dimensions in ADHD and normal students. *Open Journal of Medical Psychology*, 5, 88–91. DOI: 10.4236/ojmp.2016.54010. 82

Amado-Caballero, P., Casaseca-de-la-Higuera, P., Alberola-Lopez, S., Andres-de-Llano, J. M., Lopez-Villalobos, J. A., Garmendia-Leiza, J. R., and Alberola-Lopez, C. (2020). Objective

ADHD diagnosis using convolutional neural networks over daily-life activity records. *IEEE Journal of Biomedical and Health Informatics*. DOI: 10.1109/JBHI.2020.2964072. 30

American Psychiatric Association. (1978). *Diagnostic and Statistical Manual of Mental Disorders*. 2. Arlington, VA: American Psychiatric Association. 2

American Psychiatric Association. (1980). *Diagnostic and Statistical Manual of Mental Disorders*. 3. Arlington, VA: American Psychiatric Association. 2

American Psychiatric Association. (1994). *Diagnostic and Statistical Manual of Mental Disorders*. 4. Arlington, VA: American Psychiatric Association. 3

American Psychiatric Association. (2013). *Diagnostic and Statistical Manual of Mental Disorders* (DSM-5®). American Psychiatric Pub. DOI: 10.1176/appi.books.9780890425596. 3

Anastopoulos, A. D., Smith, T. F., Garrett, M. E., Morrissey-Kane, E., Schatz, N. K., Sommer, J. L., Kollins, S. H., and Ashley-Koch, A. (2011). Self-regulation of emotion, functional impairment, and comorbidity among children with AD/HD. *Journal of Attention Disorders*, 15(7), 583–592. DOI: 10.1177/1087054710370567. 45

Andrade, L. C. V. D., Carvalho, L. A. V., Lima, C., Cruz, A., Mattos, P., Franco, C., Soares, A., and Grieco, B. (2006). Supermarket game: An adaptive intelligent computer game for attention deficit/hyperactivity disorder diagnosis. *2006 Fifth Mexican International Conference on Artificial Intelligence* (pp. 359–368). DOI: 10.1109/MICAI.2006.45. 26

Andrejevic, M. and Selwyn, N. (2020). Facial recognition technology in schools: critical questions and concerns. *Learning, Media and Technology*, 45(2), 115–128. 73

Areces, D., Dockrell, J., García, T., Gonzaález-Castro, P., and Rodríguez, C. (2018a). Analysis of cognitive and attentional profiles in children with and without ADHD using an innovative virtual reality tool. *PLoS ONE*, 13(8), 1–18. DOI: 10.1371/journal.pone.0201039. 17, 25

Areces, D., Rodríguez, C., García, T., Cueli, M., and González-Castro, P. (2018b). Efficacy of a continuous performance test based on virtual reality in the diagnosis of ADHD and its clinical presentations. *Journal of Attention Disorders*, 22(11), 1081–1091. DOI: 10.1177/1087054716629711. 17, 25

Areces, D., García, T., Cueli, M., and Rodríguez, C. (2019). Is a virtual reality test able to predict current and retrospective ADHD symptoms in adulthood and adolescence? *Brain Sciences*, 9(10). DOI: 10.3390/brainsci9100274. 17, 25

Ariyarathne, G., De Silva, S., Dayarathna, S., Meedeniya, D., and Jayarathne, S. (2020). ADHD identification using convolutional neural network with seed-based approach for FMRI data. *Proceedings of the 2020 9th International Conference on Software and Computer Applications* (pp. 31–35). DOI: 10.1145/3384544.3384552. 32

Arns, M., Clark, C. R., Trullinger, M., DeBeus, R., Mack, M., and Aniftos, M. (2020). Neurofeedback and attention-deficit/hyperactivity-disorder (ADHD) in children: Rating the evidence and proposed guidelines. *Applied Psychophysiology and Biofeedback*, 45(2), 39–48. 40, 41

Arns, M., De Ridder, S., Strehl, U., Breteler, M., and Coenen, A. (2009). Efficacy of neurofeedback treatment in ADHD: the effects on inattention, impulsivity and hyperactivity: a meta-analysis. *Clinical EEG and Neuroscience*, 40(3), 180–189. 63

Arns, M., Heinrich, H., and Strehl, U. (2014). Evaluation of neurofeedback in ADHD: the long and winding road. *Biological Psychology*, 95, 108–115. 40

Arns, M., Heinrich, H., Ros, T., Rothenberger, A., and Strehl, U. (2015). Neurofeedback in ADHD. *Frontiers in Human Neuroscience*, 9, 602. DOI: 10.3389/fnhum.2015.00602. 63

Asiry, O., Shen, H., Wyeld, T., and Balkhy, S. (2018). Extending attention span for children ADHD using an attentive visual interface. In *2018 22nd International Conference Information Visualisation* (IV) (pp. 188–193). DOI: 10.1109/iV.2018.00041. 17, 67

Avila-Pesantez, D., Rivera, L. A., Vaca-Cardenas, L., Aguayo, S., and Zuniga, L. (2018). Toward the improvement of ADHD children through augmented reality serious games: Preliminary results. *IEEE Global Engineering Education Conference, EDUCON*, 2018-April (pp. 843–848). DOI: 10.1109/EDUCON.2018.8363318. 17, 41

Babinski, D. E. and Welkie, J. (2019). Feasibility of ecological momentary assessment of negative emotion in girls with ADHD: A pilot study. *Psychological Reports*, 123(4), 1027–1043. DOI: 10.1177/0033294119838757. 17, 52, 53

Bakker, S., van den Hoven, E., and Antle, A. N., (2011). MoSo tangibles: evaluating embodied learning. In *Proceedings of the Fifth International Conference on Tangible, Embedded, and Embodied Interaction* (pp. 85–92). ACM. DOI: 10.1145/1935701.1935720. 90

Bandura, A. (1986). *Social Foundations of Thought and Action: A Social Cognitive Theory*. Upper Saddle River, NJ: Prentice Hall. 47

Barkley, R. A. (2006). *Attention-Deficit Hyperactivity Disorder. A Handbook for Diagnosis and Treatment*, 3rd ed. New York: Guilford Press. 45, 47

Barkley, R. A. (2013). Distinguishing sluggish cognitive tempo from attention deficit hyperactivity disorder in adults. *Journal of Abnormal Psychology*, 121, 978–990. DOI: 10.1037/a0023961. 82

Bartscherer, M. L., Bartscherer, M. L., and Dole, R. L. (2005). Interactive Metronome® training for a 9-year-old boy with attention and motor coordination difficulties. *Physiotherapy Theory and Practice*, 21(4), 257–269. 89

Bashiri, A., Ghazisaeedi, M., Shahmoradi, L., Alizadeh Savareh, B., Beigy, H., Rostam Niakan Kalhori, S., Nosratabadi, M., and Estaki, M. (2018). Designing a clinical decision support system for recommending computerized cognitive rehabilitation programs: The experience of attention deficit hyperactivity disorder. *2018 2nd National and 1st International Digital Games Research Conference: Trends, Technologies, and Applications, DGRC* 2018 (pp. 34–39). DOI: 10.1109/DGRC.2018.8712064. 17

Basmajian, J. V. (1979). *Biofeedback: Principles and Practice for Clinicians.* Williams and Wilkins. 62

Batra, S., Baker, R. A., Wang, T., Forma, F., DiBiasi, F., and Peters-Strickland, T. (2017). Digital health technology for use in patients with serious mental illness: a systematic review of the literature. *Medical Devices*, 10, 237. DOI: 10.2147/MDER.S144158. 5

Belanger, L. (2017). Those with ADHD might make better entrepreneurs. Here's why. *Entrepreneur*. https://www.entrepreneur.com/article/286808. 4

Benzing, V. and Schmidt, M. (2017). Cognitively and physically demanding exergaming to improve executive functions of children with attention deficit hyperactivity disorder: A randomised clinical trial. *BMC Pediatrics*, 17(1), 1–8. DOI: 10.1186/s12887-016-0757-9. 17, 42

Benzing, V. and Schmidt, M. (2019). The effect of exergaming on executive functions in children with ADHD: A randomized clinical trial. *Scandinavian Journal of Medicine and Science in Sports*, 29(8), 1243–1253. DOI: 10.1111/sms.13446. 17, 42, 94

Benzing, V., Chang, Y. K., and Schmidt, M. (2018). Acute physical activity enhances executive functions in children with ADHD. *Scientific Reports*, 8(1), 1–10. DOI: 10.1038/s41598-018-30067-8. 17, 42

Berner, E. S. (2007). *Clinical Decision Support Systems* (Vol. 233). New York: Springer Science+Business Media, LLC. 22

Bernier, A., Carlson, S. M., and Whipple, N. (2010). From external regulation to self-regulation: Early parenting precursors of young children's executive functioning. *Child Development*, 81(1), 326–339. DOI: 10.1111/j.1467-8624.2009.01397.x. 55

Berwid, O. G. and Halperin, J. M. (2012). Emerging support for a role of exercise in attention-deficit/hyperactivity disorder intervention planning. *Current Psychiatry Reports*, 14, 543–551. DOI: 10.1007/s11920-012-0297-4. 92

Best, J. R. (2010). Effects of physical activity on children's executive function: Contributions of experimental research on aerobic exercise. *Developmental Review*, 30, 331–351. DOI: 10.1016/j.dr.2010.08.001. 92

Beyens, I., Valkenburg, P. M., and Piotrowski, J. T. (2018). Screen media use and ADHD-related behaviors: Four decades of research. *Proceedings of the National Academy of Sciences*, 115(40), 9875–9881. DOI: 10.1073/pnas.1611611114. 5

Bhuyan, S. S., Kim, H., Isehunwa, O. O., Kumar, N., Bhatt, J., Wyant, D. K., Kedia, S., Chang, D. F., and Dasgupta, D. (2017). Privacy and security issues in mobile health: Current research and future directions. *Health Policy and Technology*, 6(2), 188–191. DOI: 10.1016/j.hlpt.2017.01.004. 5

Bickett, L. and Milich, R. (1987). First impressions of learning disabled and attention deficit disordered boys. In *Biennial Meeting of the Society for Research in Child Development*, Baltimore, MD. 47

Biederman, J., Mick, E. R., Aleardi, M., Potter, A., and Herzig, K. (2005). A stimulated workplace experience for non-medicated adults with and without ADHD. *Psychiatric Services*, 56, 1617–1620. DOI: 10.1176/appi.ps.56.12.1617. 82

Bijlenga, D., Ulberstad, F., Thorell, L. B., Christiansen, H., Hirsch, O., and Kooij, J. J. S. (2019). Objective assessment of ADHD in older adults compared to controls using the QbTest. *International Journal of Geriatric Psychiatry*, 34, 1526–1533. DOI: 10.1002/gps.5163. 23

Bioulac, S., Micoulaud-Franchi, J. A., Maire, J., Bouvard, M. P., Rizzo, A. A., Sagaspe, P., and Philip, P. (2018). Virtual remediation versus methylphenidate to improve distractibility in children with ADHD: A controlled randomized clinical trial study. *Journal of Attention Disorders*, 24(2), 326–335. DOI: 10.1177/1087054718759751. 42

Birnbaum, H. G., Kessler, R. C., Lowe, S. W., Secnik, K., Greenberg, P. E., Leong, S. A., and Swensen, A. R. (2005). Costs of attention deficit–hyperactivity disorder (ADHD) in the US: excess costs of persons with ADHD and their family members in 2000. *Current Medical Research and Opinion*, 21(2), 195–205. DOI: 10.1185/030079904X20303. 1

Biswas, S. D., Chakraborty, R., and Pramanik, A. (2020). A brief survey on various prediction models for detection of ADHD from brain-MRI images. In *Proceedings of the 21st International Conference on Distributed Computing and Networking* (pp. 1–5). DOI: 10.1145/3369740.3372775. 32

Blair, C. and Diamond, A. (2008). Biological processes in prevention and intervention: The promotion of self-regulation as a means of preventing school failure. *Development and Psychopathology*, 20(3), 899. DOI: 10.1017/S0954579408000436. 45

Blandon, D. Z., Munoz, J. E., Lopez, D. S., and Gallo, O. H. (2016). Influence of a BCI neuro-feedback videogame in children with ADHD. Quantifying the brain activity through an EEG signal processing dedicated toolbox. In *2016 IEEE 11th Colombian Computing Conference, CCC 2016 - Conference Proceedings* (pp. 1–8). DOI: 10.1109/ColumbianCC.2016.7750788. 41

Blume, H. (1998). Neurodiversity: On the neurological underpinnings of geekdom. *The Atlantic*. https://www.theatlantic.com/magazine/archive/1998/09/neurodiversity/305909/. 4

Bolic, V., Lidström, H., Thelin, N., Kjellberg, A., and Hemmingsson, H. (2013). Computer use in educational activities by students with ADHD. *Scandinavian Journal of Occupational Therapy*, 20(5), 357–364. DOI: 10.3109/11038128.2012.758777. 66

Boroujeni, Y. K., Rastegari, A. A., and Khodadadi, H. (2019). Diagnosis of attention deficit hyperactivity disorder using non-linear analysis of the EEG signal. *IET Systems Biology*, 13(5), 260–266. DOI: 10.1049/iet-syb.2018.5130. 33

Bouck, E. C., Maeda, Y., and Flanagan, S. M. (2012). Assistive technology and students with high-incidence disabilities: Understanding the relationship through the NLTS2. *Remedial and Special Education*, 33(5), 298–308. DOI: 10.1177/0741932511401037. 66

Boujarwah, F. A., Kim, J. G., Abowd, G. D., and Arriaga, R. I. (2011). Developing scripts to teach social skills: Can the crowd assist the author?. *AAAIWS'11-11: Proceedings of the 11th AAAI Conference on Human Computation January 2011* (pp. 84–85). 47

Boyd, L. E., Ringland, K. E., Haimson, O. L., Fernandez, H., Bistarkey, M., and Hayes, G. R. (2015). Evaluating a collaborative iPad game's impact on social relationships for children with autism spectrum disorder. *ACM Transactions on Accessible Computing* (TACCESS), 7(1), 1–18. DOI: 10.1145/2751564. 48

Bozionelos, N. and Bozionelos, G. (2013). Research briefs: Attention deficit/hyperactivity disorder at work: Does it impact job performance. *Academy of Management Perspectives*, 27(3), 1–3. DOI: 10.5465/amp.2013.0107. 82

Breider, S., de Bildt, A., Nauta, M. H., Hoekstra, P. J., and van den Hoofdakker, B. J. (2019). Self-directed or therapist-led parent training for children with attention deficit hyperactivity disorder? A randomized controlled non-inferiority pilot trial. *Internet Interventions*, 18(August). DOI: 10.1016/j.invent.2019.100262. 17, 57

Briscoe-Smith, A. M. and Hinshaw, S. P. (2006). Linkages between child abuse and attention-deficit/hyperactivity disorder in girls: Behavioral and social correlates. *Child Abuse and Neglect*, 30(11), 1239–1255. DOI: 10.1016/j.chiabu.2006.04.008. 51

Bronson, M. (2000). *Self-Regulation in Early Childhood: Nature and Nurture*. Guilford Press. 55

Brown, V. (2009). Review of research: Individuals with ADHD lost in hyperspace. *Childhood Education*, 86(1), 45–48. DOI: 10.1080/00094056.2009.10523110. 73

Bruce, C. R., Unsworth, C. A., Dillon, M. P., Tay, R., Falkmer, T., Bird, P., and Carey, L. M. (2017). Hazard perception skills of young drivers with attention deficit hyperactivity disorder (ADHD) can be improved with computer based driver training: An exploratory randomised controlled trial. *Accident Analysis and Prevention*, 109(June), 70–77. DOI: 10.1016/j.aap.2017.10.002. 17, 80

Bryant, D. P. and Bryant, B. R. (1998). Using assistive technology adaptations to include students with learning disabilities in cooperative learning activities. *Journal of Learning Disabilities*, 31, 41–54. DOI: 10.1177/002221949803100105. 71

Bucci, L. D., Chou, T. S., and Krichmar, J. L. (2014). Tactile sensory decoding in a neuromorphic interactive robot. In *IEEE Conference on Robotics and Automation*, Hong Kong. DOI: 10.1109/ICRA.2014.6907111. 29

Budman, S. H. (2000). Behavioral health care dot-com and beyond: Computer-mediated communications in mental health and substance abuse treatment. *American Psychologist*, 55(11), 1290. DOI: 10.1037/0003-066X.55.11.1290. 5

Bul, K. C. M., Doove, L. L., Franken, I. H. A., Van Der Oord, S., Kato, P. M., and Maras, A. (2018). A serious game for children with attention deficit hyperactivity disorder: Who benefits the most? *PLoS ONE*, 13(3), 1–18. DOI: 10.1371/journal.pone.0193681. 17, 48, 78

Bul, K. C., Franken, I. H., Van der Oord, S., Kato, P. M., Danckaerts, M., Vreeke, L. J., Willems, A., van Oers, J. J. J., van den Heuvel, R., van Slagmaat, R., and Maras, A. (2015). Development and user satisfaction of "Plan-It Commander," a serious game for children with ADHD. *Games for Health Journal*, 4(6), 502–512. DOI: 10.1089/g4h.2015.0021. 17, 47, 78

Bul, K. C., Kato, P. M., Van der Oord, S., Danckaerts, M., Vreeke, L. J., Willems, A., van Oers, J. J. J., van den Heuvel, R., Birnie, D., van Amelsvoort, T. A. M. J., Franken, I. H., and Maras, A. (2016). Behavioral outcome effects of serious gaming as an adjunct to treatment for children with attention-deficit/hyperactivity disorder: A randomized controlled trial. *Journal of Medical Internet Research*, 18(2), e26. DOI: 10.2196/jmir.5173. 17, 48

Bunford, N., Evans, S. W., and Langberg, J. M. (2018). Emotion dysregulation is associated with social impairment among young adolescents with ADHD. *Journal of Attention Disorders*, 22(1), 66–82. DOI: 10.1177/1087054714527793. 45

Bunzynski, T. H., Stoyva, J. M., Adler, C. S., and Mullaney, D. J. (1973). EMG biofeedback and tension headache: a controlled outcome study. *Psychosomatic Medicine*, 35(6), 484–496. DOI: 10.1097/00006842-197311000-00004. 64

Burke, R. V., Allen, K. D., Howard, M. R., Downey, D., Matz, M. G., and Bowen, S. L. (2013). Tablet-based video modeling and prompting in the workplace for individuals with autism. *Journal of Vocational Rehabilitation*, 38(1), 1–14. DOI: 10.3233/JVR-120616. 83

Bussalb, A., Congedo, M., Barthélemy, Q., Ojeda, D., Acquaviva, E., Delorme, R., and Mayaud, L. (2019). Clinical and experimental factors influencing the efficacy of Neurofeedback in ADHD: a meta-analysis. *Frontiers in Psychiatry*, 10, 35. DOI: 10.3389/fpsyt.2019.00035. 63

Butler, A. C., Chapman, J. E., Forman, E. M., and Beck, A. T. (2006). The empirical status of cognitive-behavioral therapy: a review of meta-analyses. *Clinical Psychology Review*, 26(1), 17–31. DOI: 10.1016/j.cpr.2005.07.003. 46

Bye, L. and Jussim, L. (1993). A proposed model for the acquisition of social knowledge and social competence. *Psychology in the Schools*, 30(2), 143–161. DOI: 10.1002/1520-6807(199304)30:2<143::AID-PITS2310300207>3.0.CO;2-P. 47

Byrnes, E. and Johnson, J. H. (1981). Change technology and the implementation of automation in mental health care settings. *Behavior Research Methods and Instrumentation*, 13(4), 573–580. DOI: 10.3758/BF03202067. 6

Camacho-Conde, J. A. and Climent, G. (2020). Attentional profile of adolescents with ADHD in virtual-reality dual execution tasks: A pilot study. *Applied Neuropsychology. Child*, 1–10. DOI: 10.1080/21622965.2020.1760103. 24

Cameron, L. D. and Leventhal, H. (1995). Vulnerability beliefs, symptom experiences, and the processing of health threat information: a self-regulatory perspective. *Journal of Applied Social Psychology*, 25(21), 1859–1883. DOI: 10.1111/j.1559-1816.1995.tb01821.x. 47

Caro, K., Beltrán, J., Martínez-García, A. I., and Soto-Mendoza, V. (2016). ExercaveRoom: a technological room for supporting gross and fine motor coordination of children with developmental disorders. In *PervasiveHealth* (pp. 313–317). DOI: 10.4108/eai.16-5-2016.2263860. 94

Caro, K., Tentori, M., Martinez-Garcia, A. I., and Zavala-Ibarra, I. (2017). FroggyBobby: An exergame to support children with motor problems practicing motor coordination exercises during therapeutic interventions. *Computers in Human Behavior*, 71, 479–498. DOI: 10.1016/j.chb.2015.05.055. 94

Cartwright, T. A. (2015). The ADHD explosion: Myths, medication, money, and today's push for performance. *American Medical Writers Association Journal*, 30(2), 74. 82

Chacko, A., Bedard, A. C., Marks, D. J., Feirsen, N., Uderman, J. Z., Chimiklis, A., Rajwan, E., Cornwell, M., Anderson, L., Zwilling, A., and Ramon, M. (2014). A randomized clinical trial of Cogmed Working Memory Training in school-age children with ADHD: A replication in a diverse sample using a control condition. *Journal of Child Psychology and Psychiatry and Allied Disciplines*, 55(3), 247–255. DOI: 10.1111/jcpp.12146. 68

Chacko, A., Jensen, S. A., Lowry, L. S., Cornwell, M., Chimiklis, A., Chan, E., Lee, D., and Pulgarin, B. (2016). Engagement in behavioral parent training: Review of the literature and implications for practice. *Clinical Child and Family Psychology Review*, 19(3), 204–215. DOI: 10.1007/s10567-016-0205-2. 57

Chang, Y. and Edwards, J. K. (2015). Examining the relationships among self-efficacy, coping and job satisfaction using social career cognitive theory: An SEM analysis. *Journal of Career Assessment*, 23(1), 35–47. DOI: 10.1177/1069072714523083. 82

Chatthong, W., Khemthong, S., and Wongsawat, Y. (2020). Brain mapping performance as an occupational therapy assessment aid in attention deficit hyperactivity disorder. *American Journal of Occupational Therapy*, 74(2), 7402205070p1–7402205070p7. DOI: 10.5014/ajot.2020.035477. 34

Chen, Y., Zhang, Y., Jiang, X., Zeng, X., Sun, R., and Yu, H. (2018). COSA: Contextualized and objective system to Ssupport ADHD diagnosis. *2018 IEEE International Conference on Bioinformatics and Biomedicine (BIBM)*, 1195–1202. DOI: 10.1109/BIBM.2018.8621308. 17, 28

Chity, N., Harvey, J. R., Quadri, S., Stein, S., and Pete, S. (2012). Assistive technology as a complement to the learning style of post-secondary students with ADHD: Recommendations for design. *Thinking Differently*. OCAD University, Ontario: https://www.researchgate.net/profile/Suzanne_Stein/publication/263047136_Thinking_DIfferently/links/0f317539a59ada7024000000.pdf.

Cho, M. H. (2004). *The Effects of Design Strategies for Promoting Students' Self-Regulated Learning Skills on Students' Self-Regulation and Achievements in Online Learning Environments*. Association for Educational Communications and Technology. 72

Chou, T. S., Bucci, L. D., and Krichmar, J. L. (2015). Learning touch preferences with a tactile robot using dopamine modulated STDP in a model of insular cortex. *Frontiers in Neurorobotics*, 9, 6. DOI: 10.3389/fnbot.2015.00006. 29

Chua, R. N., Hau, Y. W., Tiew, C. M., and Hau, W. L. (2019). Investigation of attention deficit/hyperactivity disorder assessment using electro interstitial scan based on chronoamperometry technique. *IEEE Access*, 7, 144679–144690. DOI: 10.1109/ACCESS.2019.2938095. 32

Cibrian, F. L., Lakes, K. D., Tavakoulnia, A., Guzman, K., Schuck, S., and Hayes, G. R. (2020a). Supporting self-regulation of children with ADHD using wearables: Tensions and design challenges. In *Proceedings of the 2020 CHI Conference on Human Factors in Computing Systems* (pp. 1–13). DOI: 10.1145/3313831.3376837. 17, 53, 60, 70, 95

Cibrian, F. L., Madrigal, M., Avelais, M., and Tentori, M. (2020b). Supporting coordination of children with ASD using neurological music therapy: A pilot randomized control trial comparing an elastic touch-display with tambourines. *Research in Developmental Disabilities*, 106. DOI: 10.1016/j.ridd.2020.103741. 94

Cibrian, F. L., Peña, O., Ortega, D., and Tentori, M. (2017). BendableSound: An elastic multisensory surface using touch-based interactions to assist children with severe autism during music therapy. *International Journal of Human-Computer Studies*, 107, 22–37. DOI: 10.1016/j.ijhcs.2017.05.003. 94

Clancy, T. A., Rucklidge, J. J., and Owen, D. (2006). Road-crossing safety in virtual reality: A comparison of adolescents with and without ADHD. *Journal of Clinical and Adolescent Psychology*, 32(2), 203–215. DOI: 10.1207/s15374424jccp3502_4. 17, 80

Clark-Turner, M. and Begum, M. (2017). Deep recurrent Q-learning of behavioral intervention delivery by a robot from demonstration data. *RO-MAN 2017 - 26th IEEE International Symposium on Robot and Human Interactive Communication*, 2017-January (pp. 1024–1029). DOI: 10.1109/ROMAN.2017.8172429. 17, 61, 78

Classi, P., Milton, D., Ward, S., Sarsour, K., and Johnston, J. (2012). Social and emotional difficulties in children with ADHD and the impact on school attendance and healthcare utilization. *Child and Adolescent Psychiatry and Mental Health*, 6(1), 33. DOI: 10.1186/1753-2000-6-33. 45

Climent, G. and Banterla, F. (2010). *AULA, Evaluación Ecológica de los Procesos Atencionales [AULA, Ecological Evaluation of Attentional Processes]*. San Sebastián: Nesplora. 24

Coetzer, G. (2016). An empirical examination of the relationship between adult attention deficit and the operational effectiveness of project managers. *International Journal of Managing Projects in Business*, 9(2), 583–605. DOI: 10.1108/IJMPB-01 -2016-0004. 77

Cole, P. M., Martin, S. E., and Dennis, T. A. (2004). Emotion regulation as a scientific construct: Methodological challenges and directions for child development research. *Child Development*, 75, 317–333. DOI: 10.1111/j.1467-8624.2004.00673.x. 52

Coleman, B., Marion, S., Rizzo, A., Turnbull, J., and Nolty, A. (2019). Virtual reality assessment of classroom – related attention: An ecologically relevant approach to evaluating the effectiveness of working memory training. *Frontiers in Psychology*, 10, 1–9. DOI: 10.3389/fpsyg.2019.01851. 24

Comer, J. S. and Myers, K. (2016). Future directions in the use of telemental health to improve the accessibility and quality of children's mental health services. *Journal of Child and Adolescent Psychopharmacology*, 26(3), 296–300. DOI: 10.1089/cap.2015.0079. 5

Connor, D. J. (2012). Helping students with disabilities transition to college: 21 tips for students with LD and/or ADD/ADHD. *Teaching Exceptional Children*, 44(5), 16–25. DOI: 10.1177/004005991204400502. 70

Corkum, P., Elik, N., Blotnicky-Gallant, P. A., McGonnell, M., and McGrath, P. (2019). Web-based intervention for teachers of elementary students with ADHD: Randomized controlled trial. *Journal of Attention Disorders*, 23(3), 257–269. DOI: 10.1177/1087054715603198. 17, 38

Cortese, S. and Castellanos, F. X. (2012). Neuroimaging of attention-deficit/hyperactivity disorder: current neuroscience-informed perspectives for clinicians. *Current Psychiatry Reports*, 14(5), 568–578. DOI: 10.1007/s11920-012-0310-y. 33

Cortese, S., Ferrin, M., Brandeis, D., Buitelaar, J., Daley, D., Dittmann, R. W., Holtmann, M., Santosh, P., Stevenson, J., Stringaris, A., Zuddas, A., Sonuga-Barke, E. J. S., and European ADHD Guidelines Group (EAGG). (2015). Cognitive training for attention-deficit/hyperactivity disorder: Meta-analysis of clinical and neuropsychological outcomes from randomized controlled trials. *Journal of the American Academy of Child and Adolescent Psychiatry*, 54(3), 164–174. DOI: 10.1016/j.jaac.2014.12.010. 39, 40, 43

Cortese, S., Ferrin, M., Brandeis, D., Holtmann, M., Aggensteiner, P., Daley, D., Santosh, P., Simonoff, E., Stevenson, J., Stringaris, A., Sonuga-Barke, E. J., and European ADHA Guidelines Group (EAGG) (2016). Neurofeedback for attention-deficit/hyperactivity disorder: meta-analysis of clinical and neuropsychological outcomes from randomized controlled trials. *Journal of the American Academy of Child and Adolescent Psychiatry*, 55(6), 444–455. DOI: 10.1016/j.jaac.2016.03.007. 41, 63

Craven, M. P., Young, Z., Simons, L., Schnädelbach, H., and Gillott, A. (2014). From snappy app to screens in the wild: Gamifying an attention deficit hyperactivity disorder continuous performance test for public engagement and awareness. In *2014 International Conference on Interactive Technologies and Games* (pp. 36-43). IEEE. DOI: 10.1109/iTAG.2014.12. 25

Crawford, S. G., Kaplan, B. J., and Dewey, D. (2006). Effects of coexisting disorders on cognition and behavior in children with ADHD. *Journal of Attention Disorders*, 10(2), 192–199. DOI: 10.1177/1087054706289924. 88

Crowe, B. J. and Rio, R. (2004). Implications of technology in music therapy practice and research for music therapy education: A review of literature. *Journal of Music Therapy*, 41(4), 282–320. DOI: 10.1093/jmt/41.4.282. 48

Dabkowska, M. M., Pracka, D., and Pracki, T. (2007). Does actigraphy differentiate ADHD subtypes? *European Psychiatry*, 22(S1), S319–S319. DOI: 10.1016/j.eurpsy.2007.01.1062. 30

Dahlin, K. I. E. (2011). Effects of working memory training on reading in children with special needs. *Reading and Writing*, 24(4), 479–491. DOI: 10.1007/s11145-010-9238-y. 68

Danforth, J. S., Harvey, E., Ulaszek, W. R., and McKee, T. E. (2006). The outcome of group parent training for families of children with attention-deficit hyperactivity disorder and defiant/aggressive behavior. *Journal of Behavior Therapy and Experimental Psychiatry*, 37(3), 188–205. DOI: 10.1016/j.jbtep.2005.05.009. 56

Dautenhahn, K., Nehaniv, C. L., Walters, M. L., Robins, B., Kose-Bagci, H., Assif, N., and Blow, M. (2009). KASPAR–a minimally expressive humanoid robot for human–robot interaction research. *Applied Bionics and Biomechanics*, 6(3, 4), 369–397. DOI: 10.1155/2009/708594. 49

Davidson, M. C., Amso, D., Anderson, L. C., and Diamond, A. (2006). Development of cognitive control and executive functions from 4 to 13 years: Evidence from manipulations of memory, inhibition, and task switching. *Neuropsychologica*, 44, 2037–2078. DOI: 10.1016/j.neuropsychologia.2006.02.006. 37

Davis, N. O., Bower, J., and Kollins, S. H. (2018). Proof-of-concept study of an at-home, engaging, digital intervention for pediatric ADHD. *PLoS ONE*, 13(1). DOI: 10.1371/journal.pone.0189749. 18, 38

de Greef, K., Van der Spek, E. D., and Bekker, T. (2013). Designing Kinect games to train motor skills for mixed ability players. In *Games for Health* (pp. 197-205). Springer Vieweg, Wiesbaden. DOI: 10.1007/978-3-658-02897-8_15. 95

Deitz, J. C., Kartin, D., and Kopp, K. (2007). Review of the Bruininks-Oseretsky test of motor proficiency, (BOT-2). *Physical and Occupational Therapy in Pediatrics*, 27(4), 87–102. DOI: 10.1080/J006v27n04_06. 87

Denckla, M. B. (1985). Revised neurological examination for subtle signs. *Psychopharmacol Bulletin*, 21, 773–800. 87

Denckla, M. B. and Rudel, R. G. (1978). Anomalies of motor development in hyperactive boys. *Annals of Neurology: Official Journal of the American Neurological Association and the Child Neurology Society*, 3(3), 231–233. DOI: 10.1002/ana.410030308. 87

Diamond, A. (2000). Close interrelation of motor development and cognitive development and of the cerebellum and prefrontal cortex. *Child Development*, 71, 44–56. DOI: 10.1111/1467-8624.00117. 87

Diamond, A. (2015). Effects of physical exercise on executive functions: Going beyond simply moving to moving with thought. *Annals of Sports Medicine and Research*, 2(1), 1011–1016. 91, 92

Diamond, A. and Lee, K. (2011). Interventions shown to aid executive function development in children 4 to 12 years old. *Science*, 333, 959–964. DOI: 10.1126/science.1204529. 92

Díaz-Orueta, U., Garcia-López, C., Crespo-Eguílaz, N., Sánchez-Carpintero, R., Climent, G., and Narbona, J. (2014). AULA virtual reality test as an attention measure: Convergent validity with Conners' Continuous Performance Test. *Child Neuropsychology: A Journal on Normal and Abnormal Development in Childhood and Adolescence*, 20(3), 328–342. DOI: 10.1080/09297049.2013.792332. 18, 24, 25

Dibia, V. (2016). Foqus: A smartwatch application for individuals with adhd and mental health challenges. In *Proceedings of the 18th International ACM SIGACCESS Conference on Computers and Accessibility* (pp. 311–312). DOI: 10.1145/2982142.2982207. 78

Dipeolu, A., Hargrave, S., and Storlie, C. A. (2015). Enhancing ADHD and LD diagnostic accuracy using career instruments. *Journal of Career Development*, 42, 19–32. DOI: 10.1176/0894845314521691. 85

Dishman, E., Matthews, J., and Dunbar-Jacob, J. (2004). Everyday health: Technology for adaptive aging. In *Technology for Adaptive Aging*. National Academies Press (US). 52

Doan, M., Cibrian, F. L., Jang, A., Khare, N., Chang, S., Li, A., Schuck, S., Lakes, K. D., and Hayes, G. R. (2020, April). CoolCraig: A smart watch/phone application Ssupporting co-regulation of children with ADHD. In *Extended Abstracts of the 2020 CHI Conference on Human Factors in Computing Systems* (pp. 1–7). DOI: 10.1145/3334480.3382991. 70, 78

Doffman, Z. (2020). Exam monitoring webcam tech meets student outrage. *Forbes*. Published April 24, 2020. Retrieved from https://www.forbes.com/sites/zakdoffman/2020/04/24/no-lockdown-exams-sorry-kids-this-creepy-webcam-tech-lets-you-sit-them-at-home/#42e01c965cc5 on August 29, 2020. 73

Doshi, J. A., Hodgkins, P., Kahle, J., Sikirica, V., Cangelosi, M. J., Setyawan, J., Erder, M. H., and Neumann, P. J. (2012). Economic impact of childhood and adult attention-deficit/hyperactivity disorder in the United States. *Journal of the American Academy of Child and Adolescent Psychiatry*, 51(10), 990–1002. DOI: 10.1016/j.jaac.2012.07.008. 1

Dovis, S., Van Der Oord, S., Wiers, R. W., and Prins, P. J. M. (2015). Improving executive functioning in children with ADHD: Training multiple executive functions within the context of a computer game. A randomized double-blind placebo controlled trial. *PLoS ONE*, 10(4), 1–30. DOI: 10.1371/journal.pone.0121651. 18

DuPaul, G. J. and Eckert, T. L. (1998). Academic interventions for students with attention deficit/hyperactivity disorder: A review of the literature. *Reading and Writing Quarterly*, 14, 59–83. DOI: 10.1080/1057356980140104. 69

DuPaul, G. J. and Stoner, G. (2003). ADHD in the schools: Assessment and intervention strategies. New York: Guilford. 69

DuPaul, G. J., Kern, L., Belk, G., Custer, B., Daffner, M., Hatfield, A., and Peek, D. (2018). Face-to-face versus online behavioral parent training for young children at risk for ADHD: Treatment Engagement and Outcomes. *Journal of Clinical Child and Adolescent Psychology*, 47(sup1), S369–S383. DOI: 10.1080/15374416.2017.1342544. 18, 57

Edebol, H., Helldin, L., and Norlander, T. (2013). Measuring adult attention deficit hyperactivity disorder using the quantified behavior test plus. *Psychological Journal*, 2, 48–62. 23

Egeland, J., Aarlien, A. K., and Saunes, B. K. (2013). Few effects of far transfer of working memory training in ADHD: A randomized controlled trial. *PLoS ONE*, 8(10). DOI: 10.1371/journal.pone.0075660. 68

Elia, J., Ambrosini, P., and Berrettini, W. (2008). ADHD characteristics: I. Concurrent co-morbidity patterns in children and adolescents. *Child and Adolescent Psychiatry and Mental Health*, 2(1), 15. DOI: 10.1186/1753-2000-2-15. 46

Eliasson, A. C., Rösblad, B., and Forssberg, H. (2004). Disturbances in programming goal-directed arm movements in children with ADHD. *Developmental Medicine and Child Neurology*, 46(1), 19–27. DOI: 10.1017/S0012162204000040. 18, 89

Eloyan, A., Muschelli, J., Nebel, M. B., Liu, H., Han, F., Zhao, T., Barber, A. D., Joel, S., Pekar, J. J., Mostofsky, S. J., and Caffo, B. (2012). Automated diagnoses of attention deficit hyperactive disorder using magnetic resonance imaging. *Frontiers in Systems Neuroscience*, 6, 61. DOI: 10.3389/fnsys.2012.00061. 32

Emser, T. S., Johnston, B. A., Steele, J. D., Kooij, S., Thorell, L., and Christiansen, H. (2018). Assessing ADHD symptoms in children and adults: evaluating the role of objective measures. *Behavioral and Brain Functions*, 14(1), 11. DOI: 10.1186/s12993-018-0143-x. 23

Enriquez-Geppert, S., Smit, D., Pimenta, M. G., and Arns, M. (2019). Neurofeedback as a treatment intervention in ADHD: Current evidence and practice. *Current Psychiatry Reports*, 21(6), 46. DOI: 10.1007/s11920-019-1021-4. 40, 41, 43

Eom, H., Kim, K. K., Lee, S., Hong, Y. J., Heo, J., Kim, J. J., and Kim, E. (2019). Development of virtual reality continuous performance test utilizing social cues for children and adolescents with attention-deficit/hyperactivity disorder. *Cyberpsychology, Behavior, and Social Networking*, 22(3), 198–204. DOI: 10.1089/cyber.2018.0377. 18, 24

Epstein, J. N., Willis, M. G., Conners, C. K., and Johnson, D. E. (2000). Use of a technological prompting device to aid a student with attention deficit hyperactivity disorder to initiate and complete daily tasks: An exploratory study. *Journal of Special Education Technology*, 16(1), 19–28. DOI: 10.1177/016264340101600102. 78

Erhardt, D. and Hinshaw, S. (1994). Initial sociometric impressions of attention-deficit hyperactivity disorder and comparison boys: Predictions from social behaviors and from nonbehavioral variables. *Journal of Consulting and Clinical Psychology*, 62, 833–842. DOI: 10.1037/0022-006X.62.4.833. 46

Eslami, T. and Saeed, F. (2018). Similarity based classification of ADHD using singular value decomposition. *Proceedings of the 15th ACM International Conference on Computing Frontiers* (pp. 19–25). DOI: 10.1145/3203217.3203239. 32

Etscheidt, M. A. and Ayllon, T. (1987). Contingent exercise to decrease hyperactivity. *Journal of Child and Adolescent Psychotherapy*, 4, 192–198. 93

Evans, G. W. and Kim, P. (2013). Childhood poverty, chronic stress, self-regulation, and coping. *Child Development Perspectives*, 7(1), 43–48. DOI: 10.1111/cdep.12013. 55

Faedda, G. L., Ohashi, K., Hernandez, M., McGreenery, C. E., Grant, M. C., Baroni, A., Polcari, A., and Teicher, M. H. (2016). Actigraph measures discriminate pediatric bipolar disorder from attention-deficit/hyperactivity disorder and typically developing controls. *Journal of Child Psychology and Psychiatry*, 57(6), 706–716. DOI: 10.1111/jcpp.12520. 18

Fang, Q., Aiken, C. A., Fang, C., and Pan, Z. (2019b). Effects of exergaming on physical and cognitive functions in individuals with autism spectrum disorder: a systematic review. *Games for Health Journal*, 8(2), 74–84. DOI: 10.1089/g4h.2018.0032. 95

Fang, Y., Han, D., and Luo, H. (2019). A virtual reality application for assessment for attention deficit hyperactivity disorder in school-aged children. *Journal of Child PsycholNeuropsy-*

chiatric Disease and Treatmentogy and Psychiatry, 15, 1517–1523. DOI: 10.2147/NDT. S206742. 18

Farran, E. K., Bowler, A., Karmiloff-Smith, A., D'Souza, H., Mayall, L., and Hill, E. L. (2019). Cross-domain associations between motor ability, independent exploration, and large-scale spatial navigation; attention deficit hyperactivity disorder, williams syndrome, and typical development. *Frontiers in Human Neuroscience*, 13, 225. DOI: 10.3389/ fnhum.2019.00225. 18, 23, 25

Fields, S. A., Johnson, W. M., and Hassig, M. B. (2017). Adult ADHD: Addressing a unique set of challenges. *Journal of Family Practice*, 66(21), 68–74. 84

Firth, J., Torous, J., Carney, R., Newby, J., Cosco, T. D., Christensen, H., and Sarris, J. (2018). Digital technologies in the treatment of anxiety: recent innovations and future directions. *Current Psychiatry Reports*, 20(6), 44. DOI: 10.1007/s11920-018-0910-2. 5

Fisher, J. T. (2016). An examination of cognitive load and recall in ADHD and non-ADHD populations when viewing educational multimedia messages (Doctoral dissertation, Texas Tech University, https://ttu-ir.tdl.org/handle/2346/68077). 67

Fiske, A., Henningsen, P., and Buyx, A. (2019). Your robot therapist will see you now: ethical implications of embodied artificial intelligence in psychiatry, psychology, and psychotherapy. *Journal of Medical Internet Research*, 21(5), e13216. DOI: 10.2196/13216. 60

Fletcher, J. (2013). The effects of childhood ADHD on adult labor market outcomes. *Journal of Economic Behavior and Organization*, 46, 249–269. DOI: 10.3386/w18689. 82

Fohlmann, A. H. (2009). Social skills training [Social færdighedstræning]. In: Nordentoft, M., Melau, M., Iversen, T., and Kjær, S. Editor(s). *Psychosis in Adolescents. Symptoms, Treatment and the Future [Psykose hos Unge. Symptomer, Behandling og Fremtid]*. Copenhagen (DK): Psykiatrifondens Forlag, 161–89. 46

Fovet, F. (2007). Using distance learning electronic tools within the class to engage ADHD students: A key to inclusion? In *Proceedings - Frontiers in Education Conference, FIE* (pp. 15–20). DOI: 10.1109/FIE.2007.4417842. 71

Friedman, L. M. and Pfiffner, L. J. (2020). Behavioral interventions. In *The Clinical Guide to Assessment and Treatment of Childhood Learning and Attention Problems* (pp. 149–169). Academic Press. DOI: 10.1016/B978-0-12-815755-8.00007-1. 55

Fuermaier, A. B. M., Tucha, L., Koerts, J., Achenbrenner, S., Westermann, C., Weisbrod, M., Lange, K. W., and Tucha, O. (2013). Complex perspective memory in adults with attention deficit hyperactivity disorder. *PLoS ONE*, 8, e58338. DOI: 10.1371/journal.pone.0058338. 82

Fujiwara, C. S., Aderaldo, C. M., Raimir Filho, H., and Chaves, D. A. (2017). The internet of things as a helping tool in the daily life of adult patients with ADHD. In *GLOBECOM 2017-2017 IEEE Global Communications Conference* (pp. 1–6). IEEE. DOI: 10.1109/GLOCOM.2017.8254442. 18, 78

García-Baos, A., D'Amelio, T., Oliveira, I., Collins, P., Echevarria, C., Zapata, L. P., Liddle, E., and Supèr, H. (2019). Novel interactive eye-tracking game for training attention in children with attention-deficit/hyperactivity disorder. *Primary Care Companion to the Journal of Clinical Psychiatry*, 21(4), 1–8. DOI: 10.4088/PCC.19m02428. 18

Garcia-Zapirain, B., de la Torre Díez, I., and López-Coronado, M. (2017). Dual system for enhancing cognitive abilities of children with ADHD using leap motion and eye-tracking technologies. *Journal of Medical Systems*, 41(7). DOI: 10.1007/s10916-017-0757-9. 18

Garzotto, F. and Gonella, R. (2011). An open-ended tangible environment for disabled children's learning. In *Proceedings of the 10th International Conference on Interaction Design and Children* (pp. 52–61). DOI: 10.1145/1999030.1999037. 90

Gentry, T., Lau, S., Molinelli, A., Fallen, A., and Kriner, R. (2012). The Apple iPod Touch as a vocational support aid for adults with autism: Three case studies. *Journal of Vocational Rehabilitation*, 37(2), 75–85. DOI: 10.3233/JVR-2012-0601. 81

Gerber, P. J., Ginsberg, R., and Reiff, H. B. (1992). Identifying alterable patterns in employment success for highly successful adults with learning disabilities. *Journal of Learning Disabilities*, 25, 475–487. DOI: 10.1177/002221949202500802. 83

Ghiassian, S., Greiner, R., Jin, P., and Brown, M. R. (2016). Using functional or structural magnetic resonance images and personal characteristic data to identify ADHD and autism. *PLoS ONE*, 11(12), e0166934. DOI: 10.1371/journal.pone.0166934. 32

Giggins, O. M., Persson, U. M., and Caulfield, B. (2013). Biofeedback in rehabilitation. *Journal of Neuroengineering and Rehabilitation*, 10(1), 60. DOI: 10.1186/1743-0003-10-60. 62

Gilbert, D. L., Isaacs, K. M., Augusta, M., Macneil, L. K., and Mostofsky, S. H. (2011). Motor cortex inhibition: a marker of ADHD behavior and motor development in children. *Neurology*, 76(7), 615–621. DOI: 10.1212/WNL.0b013e31820c2ebd. 87, 88

Gillberg, C., Gillberg, I. C., Rasmussen, P., Kadesjö, B., Söderström, H., Råstam, M., Johnson, M., Rothenberger, A., and Niklasson, L. (2004). Co–existing disorders in ADHD–implications for diagnosis and intervention. *European Child and Adolescent Psychiatry*, 13(1), i80–i92. DOI: 10.1007/s00787-004-1008-4. 46

Ginsberg, Y., Beusterien, K. M., Amos, K., Jousselin, C., and Asherson, P. (2014). The unmet needs of all adults with ADHD are not the same: a focus on Europe. *Expert Review of Neurotherapeutics*, 14(7), 799–812. DOI: 10.1586/14737175.2014.926220. 77

Gisladottir, M. and Svavarsdottir, E. K. (2017). The effectiveness of therapeutic conversation intervention for caregivers of adolescents with ADHD: a quasi-experimental design. *Journal of Psychiatric and Mental Health Nursing*, 24(1), 15–27. DOI: 10.1111/jpm.12335. 56

Glueck, B. C. and Stroebel, C. F. (1975). Biofeedback and meditation in the treatment of psychiatric illnesses. *Comprehensive Psychiatry*. DOI: 10.1016/S0010-440X(75)80001-0. 64

Glynn, K. M. (2015). Predictors of postsecondary educational and employment outcomes for transition age state-federal vocational rehabilitation consumers with Attention-Deficit/Hyperactivity Disorder (ADHD) (Doctoral dissertation, The University of Texas at Austin: https://repositories.lib.utexas.edu/handle/2152/31694). 66

Goldman, T. A., Lee, F. J., and Zhu, J. (2014). Using video games to facilitate understanding of attention deficit hyperactivity disorder: a feasibility study. In *Proceedings of the First ACM SIGCHI Annual symposium on Computer-Human Interaction in Play* (pp. 115–120). DOI: 10.1145/2658537.2658707. 56

Goldstein, S. (Ed.). (2005). Editorial: Coaching as a treatment for ADHD. *Journal of Attention Disorders*, 9(2), 379–381. DOI: 10.1177/1087054705282198. 53

Graves, L., Asunda, P. A., Plant, S. J., and Goad, C. (2011). Asynchronous online access as an accommodation on students with learning disabilities and/or attention-deficit hyperactivity disorders in postsecondary STEM courses. *Journal of Postsecondary Education and Disability*, 24(4), 317–330. 73

Gray, S. A., Chaban, P., Martinussen, R., Goldberg, R., Gotlieb, H., Kronitz, R., Hockenberry, M., and Tannock, R. (2012). Effects of a computerized working memory training program on working memory, attention, and academics in adolescents with severe LD and comorbid ADHD: A randomized controlled trial. *Journal of Child Psychology and Psychiatry and Allied Disciplines*, 53(12), 1277–1284. DOI: 10.1111/j.1469-7610.2012.02592.x. 68

Grist, R., Porter, J., and Stallard, P. (2017). Mental health mobile apps for preadolescents and adolescents: a systematic review. *Journal of Medical Internet Research*, 19(5), e176. DOI: 10.2196/jmir.7332. 5

Gu, K., Kang, J., Lee, S., and Kim, K. M. (2017). Effects of interactive metronome intervention on behavior symptoms, timing, and motor function of children with ADHD. *Journal of Korean Academy of Sensory Integration*, 15(2), 35–45. DOI: 10.18064/JKASI.2017.15.2.035. 89

Gulsrud, A. C., Jahromi, L. B., and Kasari, C. (2010). The co-regulation of emotions between mothers and their children with autism. *Journal of Autism and Developmental Disorders*, 40(2), 227–237. DOI: 10.1007/s10803-009-0861-x. 56

Halaweh, M. (2020). Are universities using the right assessment tools during the pandemic and crisis times? *Higher Learning Research Communications*, 11, 1. 73

Halle, T. G. and Darling-Churchill, K. E. (2016). Review of measures of social and emotional development. *Journal of Applied Developmental Psychology*, 45, 8–18. DOI: 10.1016/j.appdev.2016.02.003. 45

Halperin, J. M., Newcorn, J. H., Matier, K., Sharma, V., McKay, K. E., and Schwartz, S. (1993). Discriminant validity of attention-deficit hyperactivity disorder. *Journal of the American Academy of Child and Adolescent Psychiatry*, 32(5), 1038–1043. DOI: 10.1097/00004583-199309000-00024. 30

Hansen, A., Broomfield, G., and Yap, M. B. (2019). A systematic review of technology-assisted parenting programs for mental health problems in youth aged 0–18 years: Applicability to underserved Australian communities. *Australian Journal of Psychology*, 71(4), 433–462. DOI: 10.1111/ajpy.12250. 5

Harlacher, J. E., Roberts, N. E., and Merrell, K. W. (2006). Classwide interventions for students with ADHD: A summary of teacher options beneficial to the whole class. *Teaching Exceptional Children*, 39, 6–12. DOI: 10.1177/004005990603900202. 65, 66

Harris, J. L. (2020). The Experience of Adults with Attention-Deficit/Hyperactivity Disorder in the Workplace (Doctoral dissertation, Walden University). 81

Hartanto, T. A., Krafft, C. E., Iosif, A. M., and Schweitzer, J. B. (2015). A trial-by-trial analysis reveals more intense physical activity is associated with better cognitive control performance in attention-deficit hyperactivity disorder. *Child Neuropsychology*. DOI: 10.1080/09297049.2-15/1-44511. 93

Hayes, G. R. and Hosaflook, S. W. (2014). *Technology for Transition and Postsecondary Success: Supporting Executive Function*. National Professional Resources Inc. 85

Hayes, G. R., Custodio, V. E., Haimson, O. L., Nguyen, K., Ringland, K. E., Ulgado, R. R., Waterhouse, A., and Weiner, R. (2015). Mobile video modeling for employment interviews for individuals with autism. *Journal of Vocational Rehabilitation*, 43(3), 275–287. DOI: 10.3233/JVR-150775. 82

Hayes, G. R., Yeganyan, M. T., Brubaker, J. R., O'Neal, L., and Hosaflook, S. W. (2013). Using. mobile technologies to support students in work transition programs. *Twenty-First Century*

Skills for Students with Autism. Brooke's Publishing: http://staging.star-uci.org/wp-content/uploads/2013/11/Using-Mobile-Technologies-to-Support-Students-in-Wor. 70, 85

Hechtman, L., Swanson, J. M., Sibley, M. H., Stehli, A., Owens, E. B., Mitchell, J. T., Arnold, L. E., Molina, B. S. G., Hinshaw, S. P., Jensen, P. S., Abikoff, H. B., Algorta, G. P., Howard, A. L., Hoza, B., Etocovitch, J., Houssais, S., Lakes, K. D., Nichols, J. Q., MTA Cooperative Group. (2016). Functional adult outcomes 16 years after childhood diagnosis of attention-deficit/hyperactivity disorder: MTA results. *Journal of the American Academy of Child and Adolescent Psychiatry*, 55(11), 945–952. DOI: 10.1016/j.jaac.2016.07.774. 77, 85

Hecker, L., Burns, L., Katz, L., Elkind, J., and Elkind, K. (2002). Benefits of assistive reading software for students with attention disorders. *Annals of Dyslexia*, 52(1), 243–272. DOI: 10.1007/s11881-002-0015-8. 67

Heemskerk, I., Volman, M., Ten Dam, G., and Admiraal, W. (2011). Social scripts in educational technology and inclusiveness in classroom practice. *Teachers and Teaching: Theory and Practice*, 17(1), 35–50. DOI: 10.1080/13540602.2011.538495. 47

Henderson, S. and Sugden, D. (1992) *Movement Assessment Battery for Children: Manual*. San Antonio, TX: Psychological Corporation. 87

Hernandez-Vela, A., Reyes, M., Igual, L., Moya, J., Violant, V., and Escalera, S. (2011). Adhd indicators modelling based on dynamic time warping from rgbd data: a feasibility study. In *VI CVC Workshop on the progress of Research and Development, Barcelona, Computer Vision Center* (pp. 59–62). 27

Heron, K. E., Miadich, S. A., Everhart, R. S., and Smyth, J. M. (2019). Ecological momentary assessment and related intensive longitudinal designs in family and couples research. In B. H. Fiese, M. Celano, K. Deater-Deckard, E. N. Jouriles, and M. A. Whisman (Eds.), *APA Handbooks in Psychology®. APA Handbook of Contemporary Family Psychology: Foundations, Methods, and Contemporary Issues Across the Lifespan* (pp. 369–385). American Psychological Association. DOI: 10.1037/0000099-021. 52

Heymen, S., Jones, K. R., Scarlett, Y., and Whitehead, W. E. (2003). Biofeedback treatment of constipation. *Diseases of the Colon and Rectum*, 46(9), 1208–1217. DOI: 10.1007/s10350-004-6717-8. 64

Hill, D. A., Belcher, L., Brigman, H. E., Renner, S., and Stephens, B. (2013). The Apple iPad™ as an innovative employment support for young adults with autism spectrum disorder and other developmental disabilities. *Journal of Applied Rehabilitation Counseling*, 44(1), 28–37. DOI: 10.1891/0047-2220.44.1.28. 81

Hilton, C. L., Attal, A., Best, J. R., Reistetter, T. A., Trapani, P., and Collins, D. (2015). Exergaming to improve physical and mental fitness in children and adolescents with autism spectrum disorders: Pilot study. *International Journal of Sports and Exercise Medicine*, 1(3), 1–6. DOI: 10.23937/2469-5718/1510017. 95

Hilton, C. L., Cumpata, K., Klohr, C., Gaetke, S., Artner, A., Johnson, H., and Dobbs, S. (2014). Effects of exergaming on executive function and motor skills in children with autism spectrum disorder: A pilot study. *American Journal of Occupational Therapy*, 68(1), 57–65. DOI: 10.5014/ajot.2014.008664. 95

Hirsch, O. and Christiansen, H. (2017). Factorial structure and validity of the quantified behavior test plus (Qb+@). *Assessment*, 24, 1037–1049. DOI: 10.1177/1073191116638426. 23

Hollis, C., Falconer, C. J., Martin, J. L., Whittington, C., Stockton, S., Glazebrook, C., and Davies, E. B. (2017). Annual research review: Digital health interventions for children and young people with mental health problems–a systematic and meta-review. *Journal of Child Psychology and Psychiatry*, 58(4), 474–503. DOI: 10.1111/jcpp.12663. 5

Honeybourne, V. (2019). *The Neurodiverse Workplace: An Employer's Guide to Managing and Working with Neurodivergent Employees, Clients and Customers*. London: Jessica Kingsley Publishers. 81

Hong, M., Park, B., Lee, S. M., Bahn, G. H., Kim, M. J., Park, S., Oh, I-H., and Park, H. (2020). Economic burden and disability-adjusted life years (DALYs) of attention deficit/hyperactivity disorder. *Journal of Attention Disorders*, 24(6), 823–829. DOI: 10.1177/1087054719864632. 1

Howard, J. (2018). ADHD study links teens' symptoms with digital media use CNN. https://www.cnn.com/2018/07/17/health/adhd-symptoms-digital-media-study/index.html. 67

Hoza, B. (2007). Peer functioning in children with ADHD. *Ambulatory Pediatrics*, 7(1 Suppl.), 101–106. DOI: 10.1016/j.ambp.2006.04.011. 46

Hoza, B., Mrug, S., Gerdes, A. C., Hinshaw, S. P., Bukowski, W. M., Gold, J. A., Kraemer, J. C., Pelham Jr., W. E., Wigal, T., and Arnold, L. E. (2005). What aspects of peer relationships are impaired in children with attention-deficit/ hyperactivity disorder? *Journal of Consulting and Clinical Psychology*, 73, 411–423. DOI: 10.1037/0022-006X.73.3.411. 46

Hsieh, Y. P., Yen, C. F., and Chou, W. J. (2019). Development and validation of the parental smartphone use management scale (PSUMS): Parents' perceived self-efficacy with adolescents with attention deficit hyperactivity disorder. *International Journal of Environmental Research and Public Health*, 16(8), 1423. DOI: 10.3390/ijerph16081423. 23

Huang, M. P. and Alessi, N. E. (1998). Current limitations into the application of virtual reality to mental health research. *Studies in Health Technology and Informatics*, 63-66. 5

Hubbard, J. A. and Newcomb, A. F. (1991). Initial dyadic peer interaction of attention deficit-hyperactivity disorder and normal boys. *Journal of Abnormal Child Psychology*, 19(2), 179–195. DOI: 10.1007/BF00909977. 47

Hubler, S. (2020). Keeping online testing honest? Or an Orwellian overreach?. *The New York Times*. https://www.nytimes.com/2020/05/10/us/online-testing-cheating-universities-coronavirus.html. 73

Hult, N., Kadesjö, J., Kadesjö, B., Gillberg, C., and Billstedt, E. (2018). ADHD and the QbTest: diagnostic validity of QbTest. *Journal of Attention Disorders*, 22, 1074–1080. DOI: 10.1177/1087054715595697. 23

Husain, F. (2020). Investigating current state-of-the-art applications of supportive technologies for individuals with ADHD. arXiv preprint arXiv:2005.09993. 7

Hyun, G. J., Park, J. W., Kim, J. H., Min, K. J., Lee, Y. S., Kim, S. M., and Han, D. H. (2018). Visuospatial working memory assessment using a digital tablet in adolescents with attention deficit hyperactivity disorder. *Computer Methods and Programs in Biomedicine*, 157, 137–143. DOI: 10.1016/j.cmpb.2018.01.022. 18, 28

Iacono, I., Lehmann, H., Marti, P., Robins, B., and Dautenhahn, K. (2011). Robots as social mediators for children with Autism-A preliminary analysis comparing two different robotic platforms. In *2011 IEEE International Conference on Development and Learning* (ICDL) (Vol. 2, pp. 1–6). IEEE. DOI: 10.1109/DEVLRN.2011.6037322. 51

Iriarte, Y., Diaz-Orueta, U., Cueto, E., Irazustabarrena, P., Banterla, F., and Climent, G. (2012). AULA—Advanced virtual reality tool for the assessment of attention: Normative study in Spain. *Journal of Attention Disorders*, 20(6), 542–568. DOI: 10.1177/1087054712465335. 18

Iriarte, Y., Diaz-Orueta, U., Cueto, E., Irazustabarrena, P., Banterla, F., and Climent, G. (2016). AULA-advanced virtual reality tool for the assessment of attention: Normative study in Spain. *Journal of Attention Disorders*, 20(6), 542–568. DOI: 10.1177/1087054712465335. 25

Jaiswal, S., Valstar, M. F., Gillott, A., and Daley, D. (2017). Automatic detection of ADHD and ASD from expressive behaviour in RGBD data. In *Proceedings - 12th IEEE International Conference on Automatic Face and Gesture Recognition, FG 2017 - 1st International Workshop on Adaptive Shot Learning for Gesture Understanding and Production, ASL4GUP*

2017, Biometrics in the Wild, Bwild 2017, Heteroge (pp. 762–769). IEEE. DOI: 10.1109/ FG.2017.95. 18, 27

Jasper, H. H., Solomon, P., and Bradley, C. (1938). Electroencephalographic analyses of behavior problem children. *American Journal of Psychiatry*, 95(3), 641–658. DOI: 10.1176/ ajp.95.3.641. 33

Jiang, X., Chen, Y., Huang, W., Zhang, T., Gao, C., Xing, Y., and Zheng, Y. (2020). WeDA: Designing and evaluating a scale-driven wearable diagnostic assessment system for children with ADHD. In *Proceedings of the 2020 CHI Conference on Human Factors in Computing Systems* (pp. 1–12). DOI: 10.1145/3313831.3376374. 31

Jimenez, F., Yoshikawa, T., Furuhashi, T., Kanoh, M., and Nakamura, T. (2016). Effects of collaborative learning between educational-support robots and children who potential symptoms of a development disability. *Proceedings - 2016 Joint 8th International Conference on Soft Computing and Intelligent Systems and 2016 17th International Symposium on Advanced Intelligent Systems, SCIS-ISIS* 2016 (pp. 266–270). DOI: 10.1109/SCIS-ISIS.2016.0064. 18, 68

Johnson, J. H. and Williams, T. A. (1975). The use of on-line computer technology in a mental health admitting system. *American Psychologist*, 30(3), 388. DOI: 10.1037/0003-066X.30.3.388. 4

Johnson, J. H., Giannetti, R. A., and Williams, T. A. (1976). Computers in mental health care delivery: A review of the evolution toward interventionally relevant on-line processing. *Behavior Research Methods and Instrumentation*, 8(2), 83–91. DOI: 10.3758/BF03201750. 5

Johnson, K. A., Dáibhis, A., Tobin, C. T., Acheson, R., Watchorn, A., Mulligan, A., Barry, E., Bradshaw, J. L., Gill, M., and Robertson, I. H. (2010). Right-sided spatial difficulties in ADHD demonstrated in continuous movement control. *Neuropsychologia*, 48(5), 1255–1264. DOI: 10.1016/j.*Neuropsychologia*.2009.12.026. 18, 34, 89

Kaiser, M. L., Schoemaker, M. M., Albaret, J. M., and Geuze, R. H. (2015). What is the evidence of impaired motor skills and motor control among children with attention deficit hyperactivity disorder (ADHD)? Systematic review of the literature. *Research in Developmental Disabilities*, 36, 338-357. DOI: 10.1016/j.ridd.2014.09.023. 88

Kam, H. J., Shin, Y. M., Cho, S. M., Kim, S. Y., Kim, K. W., and Park, R. W. (2010). Development of a decision support model for screening attention-deficit hyperactivity disorder with actigraphbased measurements of classroom activity. *Applied Clinical Informatics*, 1(4), 377–393. DOI: 10.4338/ACI-2010-05-RA-0033. 18, 30

Kaneko, M., Yamashita, Y., and Iramina, K. (2016). Quantitative evaluation system of soft neurological signs for children with attention deficit hyperactivity disorder. *Sensors* (Basel, Switzerland), 16(1). DOI: 10.3390/s16010116. 31

Kanellos, T., Doulgerakis, A., Georgiou, E., Bessa, M., Thomopoulos, S. C. A., Vatakis, A., del Val-Guardiola, A., and Navarra, J. (2019). User experience evaluation of the reefocus adhd management gaming system. *2019 4th International Conference on Smart and Sustainable Technologies*, SpliTech 2019, (732375). DOI: 10.23919/SpliTech.2019.8783141. 18, 42

Kato, P. M., Cole, S. W., Bradlyn, A. S., and Pollock, B. H. (2008). A video game improves behavioral outcomes in adolescents and young adults with cancer: a randomized trial. *Pediatrics*, 122(2), e305–e317. DOI: 10.1542/peds.2007-3134. 48

Katz, M. (2012). The zones of regulation: A curriculum designed to foster self-regulation and emotional control. *Attention* (pp. 7–8). https://chadd.org/wp-content/uploads/2018/06/ATTN_10_12_Zones.pdf. 74

Kelley, J. (2020). *Students Are Pushing Back Against Proctoring Surveillance Apps*. Electronic Frontier Foudation. https://www.eff.org/deeplinks/2020/09/students-are-pushing-back-against-proctoring-surveillance-apps. 73

Kemppinen, J., Korpela, J., Partners, R., Elfvengren, K., Salmisaari, T., Polkko, J., and Tuominen, M. (2013). A clinical decision support system for Aadult ADHD diagnostics process. *2013 46th Hawaii International Conference on System Sciences* (pp. 2616–2625). DOI: 10.1109/HICSS.2013.30. 22

Kern, L., DuPaul, G. J., Volpe, R. J., Sokol, N. G., Gary Lutz, J., Arbolino, L. A., Pipan, M., and VanBrakle, J. D. (2007). Multisetting assessment-based intervention for young children at risk for attention deficit hyperactivity disorder: Initial effects on academic and behavioral functioning. *School Psychology Review*, 36(2), 237–255. DOI: 10.1080/02796015.2007.12087942. 57

Keshav, N. U., Vogt-Lowell, K., Vahabzadeh, A., and Sahin, N. T. (2019). Digital attention-related augmented-reality game: Significant correlation between student game performance and validated clinical measures of attention-deficit/hyperactivity disorder (ADHD). *Children* (Basel, Switzerland), 6(6). DOI: 10.3390/children6060072. 23, 25

Khan, S. A. and Faraone, S. V. (2006). The genetics of ADHD: a literature review of 2005. *Current Psychiatry Reports*, 8(5), 393–397. DOI: 10.1007/s11920-006-0042-y. 97

Khoshnoud, S., Nazari, M. A., and Shamsi, M. (2018). Functional brain dynamic analysis of ADHD and control children using nonlinear dynamical features of EEG signals. *Journal of Integrative Neuroscience*, 17(1), 17–30. DOI: 10.3233/JIN-170033. 33

Khoshnoud, S., Shamsi, M., and Nazari, M. A. (2015). Non-linear EEG analysis in children with attention-deficit/ hyperactivity disorder during the rest condition. *2015 22nd Iranian Conference on Biomedical Engineering* (pp. 87–92). ICBME. DOI: 10.1109/ICBME.2015.7404122. 33

Kientz, J. A., Hayes, G. R., Goodwin, M. S., Gelsomini, M., and Abowd, G. D. (2019). Interactive technologies and autism. *Synthesis Lectures on Assistive, Rehabilitative, and Health-Preserving Technologies*. San Rafael, CA: Morgan and Claypool Publishers, 9(1), i–229. DOI; 10.2200/S00988ED2V01Y202002ARH013. 9, 12

Kim, S., Ryu, J., Choi, Y., Kang, Y., Li, H., and Kim, K. (2020). Eye-contact game using mixed reality for the treatment of children with attention deficit hyperactivity disorder. *IEEE Access*, 8, 45996–46006. DOI: 10.1109/ACCESS.2020.2977688. 18, 38

Klein, R. G., Mannuzza, S., and Olazagasti, M. A. (2012). Clinical and functional outcome of childhood attention-deficit/hyperactivity disorder 33 years later. *Archive of General Psychiatry*, 69, 1295–1303. DOI: 10.1001/archgenpsychiatry.2012.271. 82

Knight, M. (2019). Organization solutions for people with ADHD. Presented at the *National Association of Productivity and Organizing Professionals* (NAPO) South Florida Chapter Meeting at Nova Southeastern University in Fort Lauderdale, FL. 65

Knight, S. (1995). The NHS information management and technology strategy from a mental health perspective. *Advances in Psychiatric Treatment*, 1(8), 223–229. DOI: 10.1192/apt.1.8.223. 5

Knouse, L. E., Bagwell, C. L., Barkley, R. A., and Murphy, K. R. (2005). Accuracy of self-evaluation in adults with ADHD: evidence from a driving study. *Journal of Attention Disorders*, 8(4), 221–234. DOI: 10.1177/1087054705280159. 23

Kochanska, G., Coy, K. C., and Murray, K. T. (2001). The development of self-regulation in the first four years of life. *Child Development*, 72(4), 1091–1111. DOI: 10.1111/1467-8624.00336. 55

Kollins, S. H., DeLoss, D. J., Cañadas, E., Lutz, J., Findling, R. L., Keefe, R. S., Epstein, J. N., Cutler, A. J., and Faraone, S. V. (2020). A novel digital intervention for actively reducing severity of paediatric ADHD (STARS-ADHD): A randomised controlled trial. *The Lancet Digital Health*, 2(4), E168-E178. DOI: 10.1016/S2589-7500(20)30017-0. 18, 40

Konold, T. R. and Pianta, R. C. (2005). Empirically-derived, person-oriented patterns of school readiness in typically-developing children: Description and prediction to first-grade achievement. *Applied Developmental Science*, 9(4), 174–187. DOI: 10.1207/s1532480xads0904_1. 45

Kooistra, L., Crawford, S., Dewey, D., Cantell, M., and Kaplan, B. J. (2005). Motor correlates of ADHD: contribution of reading disability and oppositional defiant disorder. *Journal of Learning Disabilities*, 38(3), 195–206. DOI: 10.1177/00222194050380030201. 88

Kretzschmar, K., Tyroll, H., Pavarini, G., Manzini, A., Singh, I., and NeurOx Young People's Advisory Group. (2019). Can your phone be your therapist? Young people's ethical perspectives on the use of fully automated conversational agents (chatbots) in mental health support. *Biomedical Informatics Insights*, 11. DOI: 10.1177/117822261982908. 60

Krichmar, J. L. and Chou, T. S. (2018). A tactile robot for developmental disorder therapy. In *Proceedings of Technology, Mind, and Society* (pp. 1–6). DOI: 10.1145/3183654.3183657. 29

Kuang, D. and He, L. (2014). Classification on ADHD with deep learning. In *2014 International Conference on Cloud Computing and Big Data* (pp. 27–32). IEEE. DOI: 10.1109/CCBD.2014.42. 32

Kubik, J. A. (2010). Efficacy of ADHD coaching for adults with ADHD. *Journal of Attention Disorders*, 13(5), 442–453. DOI: 10.1177/1087054708329960. 83

Kushlev, K., Proulx, J., and Dunn, E. W. (2016, May). " Silence your phones" smartphone notifications increase inattention and hyperactivity symptoms. In *Proceedings of the 2016 CHI Conference on Human Factors in Computing Systems* (pp. 1011–1020). DOI: 10.1145/2858036.2858359. 67

Lakes, K. D., Kettler, R. J., Schmidt, J., Haynes, M., Feeney-Kettler, K., Kamptner, L., Swanson, J., and Tamm, L. (2009). The CUIDAR early intervention parent training program for preschoolers at risk for behavioral disorders: An innovative practice for reducing disparities in access to service. *Journal of Early Intervention*, 31(2), 167–178. DOI: 10.1177/1053815109331861. 78

Lakes, K. D., Vargas, D., Riggs, M., Schmidt, J., and Baird, M. (2011). Parenting intervention to reduce attention and behavior difficulties in preschoolers: A CUIDAR evaluation study. *Journal of Child and Family Studies*, 20(5), 648–659. DOI: 10.1007/s10826-010-9440-1. 78

Lakes, K., Cibrian, F., Schuck, S., Nelson, M., and Hayes, G.R. (Under Review) Digital health interventions for youth with ADHD: A systematic review. 38

Lanyon, R. I. (1971). Mental health technology. *American Psychologist*, 26(12), 1071. DOI: 10.1037/h0032233. 60

Lazarus, B. D. (2008). Accommodating persons with disabilities in virtual workplaces. In *Handbook of Research on Virtual Workplaces and the New Nature of Business Practices* (pp. 196–205). IGI Global. DOI: 10.4018/978-1-59904-893-2.ch015. 85

Le, H. H., Hodgkins, P., Postma, M. J., Kahle, J., Sikirica, V., Setyawan, J., Erder, J., and Doshi, J. A. (2014). Economic impact of childhood/adolescent ADHD in a European setting: the Netherlands as a reference case. *European Child and Adolescent Psychiatry*, 23(7), 587–598. DOI: 10.1007/s00787-013-0477-8. 1

Lehmann, H., Iacono, I., Robins, B., Marti, P., and Dautenhahn, K. (2011). "Make it move": Playing cause and effect games with a robot companion for children with cognitive disabilities. *ECCE 2011 - European Conference on Cognitive Ergonomics 2011: 29th Annual Conference of the European Association of Cognitive Ergonomics*, (August) (pp. 105–112). DOI: 10.1145/2074712.2074734. 49

Lehrer, P. M., Vaschillo, E., Vaschillo, B., Lu, S. E., Scardella, A., Siddique, M., and Habib, R. H. (2004). Biofeedback treatment for asthma. *Chest*, 126(2), 352–361. DOI: 10.1378/chest.126.2.352. 64

Lenartowicz, A. and Loo, S. K. (2014). Use of EEG to diagnose ADHD. *Current Psychiatry Reports*, 16(11), 498. DOI: 10.1007/s11920-014-0498-0. 33

Lewis, P., Noble, S., and Soiffer, N. (2010). Using accessible math textbooks with students who have learning disabilities. In *ASSETS'10 - Proceedings of the 12th International ACM SIGACCESS Conference on Computers and Accessibility* (pp. 139–146). DOI: 10.1145/1878803.1878829. 18, 69

Li, J. J. and Lansford, J. E. (2018). A smartphone-based ecological momentary assessment of parental behavioral consistency: Associations with parental stress and child ADHD symptoms. *Developmental Psychology*, 54(6), 1086. DOI: 10.1037/dev0000516. 23

Liberman, R. P. (1988). Social skills training. In: Liberman RP editor(s). *Psychiatric Rehabilitation of Chronic Mental Patients*. Washington DC: American Psychiatric Press. 46

Linden, M., Habib, T., and Radojevic, V. (1996). A controlled study of the effects of EEG biofeedback on cognition and behavior of children with attention deficit disorder and learning disabilities. *Biofeedback and Self-regulation*, 21(1), 35–49. DOI: 10.1007/BF02214148. 63

Lindstedt, H. and Umb-Carlsson, Õ. (2013). Cognitive assistive technology and professional support in everyday life for adults with ADHD. *Disability and Rehabilitation: Assistive Technology*, 8(5), 402–408. DOI: 10.3109/17483107.2013.769120. 79

Lis, S., Baer, N., Stein-en-Nosse, C., Gallhofer, B., Sammer, G., and Kirsch, P. (2010). Objective measurement of motor activity during cognitive performance in adults with attention-deficit/hyperactivity disorder. *Acta Psychiatrica Scandinavica*, 122, 285–294. DOI: 10.1111/j.1600-0447.2010.01549.x. 23, 28

Littlefield, L., Cavanagh, S., Knapp, R., and O'Grady, L. (2017). KidsMatter: Building the capacity of Australian primary schools and early childhood services to foster children's social and emotional skills and promote children's mental health. In *Social and Emotional Learning in Australia and the Asia-Pacific* (pp. 293–311). Springer, Singapore. DOI: 10.1007/978-981-10-3394-0_16. 46

Lobo, J., Matsuda, S., Futamata, I., Sakuta, R., and Suzuki, K. (2019). Chimelight: Augmenting instruments in interactive music therapy for children with neurodevelopmental disorders. *ASSETS 2019 - 21st International ACM SIGACCESS Conference on Computers and Accessibility* (pp. 124–135). DOI: 10.1145/3308561.3353784. 48

Lofthouse, N., Arnold, L. E., Hersch, S., Hurt, E., and DeBeus, R. (2012). A review of neurofeedback treatment for pediatric ADHD. *Journal of Attention Disorders*, 16(5), 351–372. DOI: 10.1177/1087054711427530. 63

Lombardi, A., Izzo, M. V., Gelbar, N., Murray, A., Buck, A., Johnson, V., Hsiao, J., Wei, Y., and Kowitt, J. (2017). Leveraging information technology literacy to enhance college and career readiness for secondary students with disabilities. *Journal of Vocational Rehabilitation*, 46(3), 389–397. DOI: 10.3233/JVR-170875. 81

Loo, S. K. and Makeig, S. (2012). Clinical utility of EEG in attention-deficit/hyperactivity disorder: a research update. *Neurotherapeutics*, 9(3), 569–587. DOI: 10.1007/s13311-012-0131-z. 33, 63

Lopez, P. L., Torrente, F. M., Ciapponi, A., Lischinsky, A. G., Cetkovich-Bakmas, M., Rojas, J. L., Romano, M., and Manes, F. F. (2013). Cognitive-behavioral interventions for attention deficit hyperactivity disorder (ADHD) in adults (protocol). *Cochrane Database of Systematic Review*, 11, 16–31. DOI: 10.1002/14651858.CD010840. 82

Lopresti, E. F.,, Mihailidis, A., and Kirsch, N. (2004). Assistive technology for cognitive rehabilitation: State of the art. Neuropsychological rehabilitation, 14(1-2), 5–39. DOI: 10.1080/09602010343000101. 79

Loren, R. E., Vaughn, A. J., Langberg, J. M., Cyran, J. E., Proano-Raps, T., Smolyansky, B. H., Tamm, L., and Epstein, J. N. (2015). Effects of an 8-session behavioral parent training group for parents of children with ADHD on child impairment and parenting confidence. *Journal of Attention Disorders*, 19(2), 158–166. DOI: 10.1177/1087054713484175. 56

LoSardo, A. (2020). Faceoff: The fight for privacy in American public schools in the wake of facial recognition technology. *Seton Hall Legislative Journal*, 44, 373. 73

Loskutova, N. Y., Callen, E., Pinckney, R. G., Staton, E. W., and Pace, W. D. (2019). Feasibility, implementation and outcomes of tablet-based two-step screening for adult ADHD in primary care practice. *Journal of Attention Disorders*. DOI: 10.1177/1087054719841133. 28

Luna, J., Treacy, R., Hasegawa, T., Campbell, A., and Mangina, E. (2018). Words worth learning-augmented literacy content for ADHD students. *2018 IEEE Games, Entertainment, Media Conference, GEM 2018* (pp. 181–188). DOI: 10.1109/GEM.2018.8516483. 18, 68

Luxton, D. D., McCann, R. A., Bush, N. E., Mishkind, M. C., and Reger, G. M. (2011). mHealth for mental health: Integrating smartphone technology in behavioral healthcare. *Professional Psychology: Research and Practice*, 42(6), 505. DOI: 10.1037/a0024485. 60

Maarek, A. (2012). Electro interstitial scan system: assessment of 10 years of research and development. *Medical Devices* (Auckland, NZ), 5, 23. DOI: 10.2147/MDER.S29319. 32

Mackenzie, A. and Nickerson, P. (2009). *The Time Trap: The Classic Book on Time Management*. Amacom. 65

Mahoney, J. L., Durlak, J. A., and Weissberg, R. P. (2018). An update on social and emotional learning outcome research. *Phi Delta Kappan*, 100(4), 18–23. DOI: 10.1177/0031721718815668. 46

Mancini, V., Rudaizky, D., Howlett, S., Elizabeth-Price, J., and Chen, W. (2020). Movement difficulties in children with ADHD: Comparing the long-and short-form Bruininks–Oseretsky Test of Motor Proficiency—Second Edition (BOT-2). *Australian Occupational Therapy Journal*, 67(2), 153–161. DOI: 10.1111/1440-1630.12641. 87

Marcano, J. L. L., Bell, M. A., and Beex, A. L. (2016). EEG channel selection for AR model based ADHD classification. In *2016 IEEE Signal Processing in Medicine and Biology Symposium (SPMB)* (pp. 1–6). IEEE. DOI: 10.1109/SPMB.2016.7846851. 33

Marti, P. (2010). Bringing playfulness to disability. In *Proceedings of the 6th Nordic Conference on Human Computer Interaction, NordiCHI 2010* (Reykjavik, Iceland, October 2010). DOI: 10.1145/1868914.1869046. 50

Marti, P., Giusti, L., and Rullo, A. (2009). Robot as social mediators: Field trials with children with special needs. In: Emliani, P. L., Burzagli, A., Como, F., Gabbanini, F., Salminen, A. (Ed.), *Assistive Technology for Adapted Equipment to Inclusive Environments, Proceedings of the AAATE 2009*, August 31–September 2, 2009, (pp. 165–169), Amsterdam: IOS Press. 50

Martinez-Raga, J., Szerman, N., Knecht, C., and Alvaro., R. (2013). Attention deficit hyperactivity disorder and dual disorders. Educational needs for an underdiagnosed condition. *International Journal of Adolescent Medicine and Health*, 25, 231–243. DOI: 10.1515/ijamh-2013-0057. 82

Marzbani, H., Marateb, H. R., and Mansourian, M. (2016). Neurofeedback: a comprehensive review on system design, methodology and clinical applications. *Basic and Clinical Neuroscience*, 7(2), 143. DOI: 10.15412/J.BCN.03070208. 63

Matic, A., Hayes, G. R., Tentori, M., Abdullah, M., and Schuck, S. (2014). Collective use of a situated display to encourage positive behaviors in children with behavioral challenges. In *UbiComp 2014 - Proceedings of the 2014 ACM International Joint Conference on Pervasive and Ubiquitous Computing* (pp. 885–895). DOI: 10.1145/2632048.2632070. 18, 59

Maurizio, S., Liechti, M. D., Heinrich, H., Jäncke, L., Steinhausen, H. C., Walitza, S., Brandeis, D., and Drechsler, R. (2014). Comparing tomographic EEG neurofeedback and EMG biofeedback in children with attention-deficit/hyperactivity disorder. *Biological Psychology*, 95, 31–44. DOI: 10.1016/j.biopsycho.2013.10.008. 63

Mautone, J. A., DuPaul, G. J., and Jitendra, A. K. (2005). The effects of computer-assisted instruction on the mathematics performance and classroom behavior of children with ADHD. *Journal of Attention Disorders*, 9(1), 301–312. DOI: 10.1177/1087054705278832. 18, 69

McClelland, M. M. and Cameron, C. E. (2012). Self-regulation in early childhood: Improving conceptual clarity and developing ecologically valid measures. *Child Development Perspectives*, 6(2), 136–142. DOI: 10.1111/j.1750-8606.2011.00191.x. 55

McClelland, M., Geldhof, J., Morrison, F., Gestsdóttir, S., Cameron, C., Bowers, E., Duckworth, A., Little, T., and Grammer, J. (2018). Self-regulation. In *Handbook of Life Course Health Development* (pp. 275–298). Springer, Cham. DOI: 10.1007/978-3-319-47143-3_12. 55

McEwan, F., Thompson, M., Laver-Bradbury, C., Jefferson, H., Koerting, J., Smith, E., Knowles, M., McCann, D., Daley, D., Barton, J., Latter, S., Elsey, H., and Sonuga-Barke, E. (2015). Innovations in practice: adapting a specialized ADHD parenting programme for use with 'hard to reach'and 'difficult to treat'preschool children. *Child and Adolescent Mental Health*, 20(3), 175–178. DOI: 10.1111/camh.12069. 57

McQuade, J. D. and Hoza, B. (2008). Peer problems in attention deficit hyperactivity disorder: Current status and future directions. *Developmental Disabilities Research Reviews*, 14, 320–324. DOI: 10.1002/ddrr.35. 45

McRae, M. (2018). There's a worrying link between rising ADHD symptoms and too much internet, study shows. *Science Alert*. https://www.sciencealert.com/attention-deficit-hyperactivity-disorder-correlation-social-media-use-adolescents. 67

Melby-Lervåg, M. and Hulme, C. (2013). Is working memory training effective? A meta-analytic review. *Developmental Psychology*, 49(2), 270. DOI: 10.1037/a0028228. 39

Meyers, C. A. and Bagnall, R. G. (2015). A case study of an adult learner with ASD and ADHD in an undergraduate online learning environment. *Australasian Journal of Educational Technology*, 31(2). DOI: 10.14742/ajet.1600. 73

Miao, B., Zhang, L. L., Guan, J. L., Meng, Q. F., and Zhang, Y. L. (2019). Classification of ADHD individuals and neurotypicals using reliable RELIEF: A resting-state study. *IEEE Access*, 7, 62163–62171. DOI: 10.1109/ACCESS.2019.2915988. 32

Michael, D. R. and Chen, S. L. (2005). *Serious Games: Games that Educate, Train, and Inform*. Muska and Lipman/Premier-Trade. 47

Michel, T., Slovak, P., and Fitzpatrick, G. (2019). An explorative review of youth mental health apps for prevention and promotion. In *13th EAI International Conference on Pervasive Computing Technologies for Healthcare-Demos and Posters. European Alliance for Innovation (EAI)*. DOI: 10.4108/eai.20-5-2019.2283578. 5

Michelsen, G., Slettebø, T., and Moser, I. B. (2017). Introduction of cognitive support technologies (CST) for job seekers. *The Journal on Technology and Persons with Disabilities*, 5(22). 80

Michelsen, G., Slettebø, T., and Moser, I. B. (2019a). Inclusive physical and digital spaces in vocational rehabilitation. *Nordic Journal of Science and Technology Studies*, 7(1), 32–41. DOI: 10.5324/njsts.v7i1.2796. 81

Michelsen, G., Slettebø, T., and Moser, I. B. (2019b). The empowering value of introducing CST in vocational rehabilitation. *Disability and Rehabilitation: Assistive Technology*, 15(2), 157–165. DOI: 10.1080/17483107.2018.1545263. 81

Micoulaud-Franchi, J. A., Geoffroy, P. A., Fond, G., Lopez, R., Bioulac, S., and Philip, P. (2014). EEG neurofeedback treatments in children with ADHD: an updated meta-analysis of randomized controlled trials. *Frontiers in Human Neuroscience*, 8, 906. DOI: 10.3389/fnhum.2014.00906. 63

Mihailidis, A. and LoPresti, E. F. (2006). Cognitive assistive technology. In Akay, M. (Ed.) *Wiley Encyclopedia of Biomedical Engineering*. DOI: 10.1002/9780471740360.ebs1340. 79

Milham, M. P., Fair, D., Mennes, M., and Mostofsky, S. H. (2012). The ADHD-200 consortium: a model to advance the translational potential of neuroimaging in clinical neuroscience. *Frontiers in Systems Neuroscience*, 6, 62. DOI: 10.3389/fnsys.2012.00062. 32

Mitchell, J. T., Zylowska, L., and Kollins, S. H. (2015). Mindfulness meditation training for attention- deficit/hyperactivity disorder in adulthood: Current empirical support, treatment overview, and future directions. *Cognitive and Behavioral Practice*, 22, 172–191. DOI: 10.1016/j.cbpra.2014.10.002. 93

Mock, P., Tibus, M., Ehlis, A. C., Baayen, H., and Gerjets, P. (2018). Predicting ADHD risk from touch interaction data. *ICMI 2018 - Proceedings of the 2018 International Conference on Multimodal Interaction* (pp. 446–454). DOI: 10.1145/3242969.3242986. 18, 25

Moffitt, T. E., Arseneault, L., Belsky, D., Dickson, N., Hancox, R. J., Harrington, H., Houts, R., Poulton, R., Roberts, B. W., Ross, S., Sears, M. R., Thomson, W. M., and Caspi, A. (2011). A gradient of childhood self-control predicts health, wealth, and public safety. *Proceedings of the National Academy of Sciences*, 108(7), 2693–2698. DOI: 10.1073/pnas.1010076108. 3, 37, 55

Mohammadi, M. R., Khaleghi, A., Nasrabadi, A. M., Rafieivand, S., Begol, M., and Zarafshan, H. (2016). EEG classification of ADHD and normal children using non-linear features and neural network. *Biomedical Engineering Letters*, 6(2), 66–73. DOI: 10.1007/s13534-016-0218-2. 33

Morash-Macneil, V., Johnson, F., and Ryan, J. B. (2018). A systematic review of assistive technology for individuals with intellectual disability in the workplace. *Journal of Special Education Technology*, 33(1), 15–26. DOI: 10.1177/0162643417729166. 85

Morgensterns, E., Alfredsson, J., and Hirvikoski, T. (2015). Structured skills training for adults with ADHD in an inpatient psychiatric context: An open feasibility trail. *Attention Deficit Hyperactivity Disorder*, 8, 101–111. DOI: 10.1007/s12402-015-0182-1. 85

Morris, M. R., Begel, A., and Wiedermann, B. (2015). Understanding the challenges faced by neurodiverse software engineering employees: Toward a more inclusive and productive technical workforce. In *Proceedings of the 17th International ACM SIGACCESS Conference on Computers and Accessibility* (pp. 173–184). DOI: 10.1145/2700648.2809841. 82

Mueller, A., Candrian, G., Kropotov, J. D., Ponomarev, V. A., and Baschera, G. M. (2010). Classification of ADHD patients on the basis of independent ERP components using a machine learning system. In *Nonlinear Biomedical Physics*, 4(S1), p. S1. BioMed Central. DOI: 10.1186/1753-4631-4-S1-S1. 33

Mull, C. A. and Sitlington, P. L. (2003). The role of technology in the transition to postsecondary education of students with learning disabilities: A review of the literature. *The Journal of Special Education*, 37(1), 26–32. DOI: 10.1177/00224669030370010301. 71

Munizza, C., Tibaldi, G., Cesano, S., Dazzi, R., Fantini, G., Palazzi, C., and Zuccolin, M. (2000). Mental health care in Piedmont: a description of its structure and components using a new technology for service assessment. *Acta Psychiatrica Scandinavica*, 102, 47–58. DOI: 10.1111/j.0902-4441.2000.acp28-06.x. 5

Muñoz-Organero, M., Powell, L., Heller, B., Harpin, V., and Parker, J. (2018). Automatic extraction and detection of characteristic movement patterns in children with ADHD based on a convolutional neural network (CNN) and acceleration images. *Sensors* (Switzerland), 18(11). DOI: 10.3390/s18113924. 18, 30

Muñoz-Organero, M., Powell, L., Heller, B., Harpin, V., and Parker, J. (2019). Using recurrent neural networks to compare movement patterns in ADHD and normally developing children based on acceleration signals from the wrist and ankle. *Sensors* (Basel, Switzerland), 19(13). DOI: 10.3390/s19132935. 18, 30

Murphy, K. (2005). Psychosocial treatments for ADHD in teens and adults: A practice-friendly review. *Journal of Clinical Psychology*, 61(5), 607–619. DOI: 10.1002/jclp.20123. 66

Murphy, K., Ratey, N., Maynard, S., Sussman, S., and Wright, S. D. (2010). Coaching for ADHD. *Journal of Attention Disorders*, 13(5), 546–552. DOI: 10.1177/1087054709344186. 53

Murray, D. W. and K Rosanbalm. (2017). Promoting Self-Regulation in Adolescents and Young Adults: A Practice Brief. OPRE Report #2015-82. Washington, DC: Office of Planning, Research, and Evaluation, Administration for Children and Families, U.S. Department of Health and Human Services (pp. 1–6). 55

Mwamba, H. M., Fourie, P. R., and van den Heever, D. (2019). PANDAS: Paediatric attention-deficit/hyperactivity disorder application software. *Applied Sciences*, 9(8), 1645. DOI: 10.3390/app9081645. 26

Nadeau, K. G. (2015). *The ADHD Guide to Career Success: Harness Your Strengths, Manage Your Challenges* (2nd ed.). New York: Taylor and Francis. DOI: 10.4324/9781315723334. 85Nakada, T., Kanai, H., and Kunifuji, S. (2005). A support system for finding lost objects using spotlight. In *Proceedings of the 7th International Conference on Human Computer Interaction with Mobile devices and Services* (pp. 321–322). DOI: 10.1145/1085777.1085846. 78

Namgung, Y., Son, D. I., and Kim, K. M. (2015). Effect of interactive metronome® training on timing, attention and motor function of children with ADHD: case report. *The Journal of Korean Academy of Sensory Integration*, 13(2), 63–73. DOI: 10.18064/JKASI.2015.13.2.063. 89

Neguț, A., Jurma, A. M., and David, D. (2017). Virtual-reality-based attention assessment of ADHD: ClinicaVR: Classroom-CPT versus a traditional continuous performance test. *Child Neuropsychology: A Journal on Normal and Abnormal Development in Childhood and Adolescence*, 23(6), 692–712. DOI: 10.1080/09297049.2016.1186617. 18, 24

Nguyen, B., Steel, P., and Ferrari, J. R. (2013). Procrastination's impact in the workplace and the workplace's impact on procrastination. *International Journal of Selection and Assessment*, 21(4), 388–399. DOI: 10.1111/ijsa.12048. 82

Nicandro, V. , Khandelwal, A., and Weitzman A. (2020). Please, let students turn their videos off in class Zoom camera usage policies are draining students, for more reasons than just Zoom fatigue. *The Standford Daily*. https://www.stanforddaily.com/2020/06/01/please-let-students-turn-their-videos-off-in-class/. 73

Nolin, P., Stipanicic, A., Henry, M., Lachapelle, Y., Lussier-Desrochers, D., Rizzo, A., and Allain, P. (2016). ClinicaVR: Classroom-CPT: A virtual reality tool for assessing attention and inhibition in children and adolescents. *Computers in Human Behavior*, 59, 327–333. DOI: 10.1016/j.chb.2016.023. 18, 24

Novick, J. M., Bunting, M. F., Engle, R. W., and Dougherty, M. R. (Eds.). (2020). *Cognitive and Working Memory Training: Perspectives from Psychology, Neuroscience, and Human Development*. New York: Oxford University Press. DOI: 10.1093/oso/9780199974467.001.0001. 39, 43

Nowacek, E. J. and Mamlin, N. (2007). General education teachers and students with ADHD: What modifications are made? *Preventing School Failure: Alternative Education for Children and Youth*, 51, 28–35. DOI: 10.3200/PSFL.51.3.28-35. 65, 66

O'Mahony, N., Florentino-Liano, B., Carballo, J. J., Baca-García, E., and Rodríguez, A. A. (2014). Objective diagnosis of ADHD using IMUs. *Medical Engineering and Physics*, 36(7), 922–926. DOI: 10.1016/j.medengphy.2014.02.023. 31

Odgers, C. L. and Jensen, M. R. (2020). Annual Research Review: Adolescent mental health in the digital age: facts, fears, and future directions. *Journal of Child Psychology and Psychiatry*, 61(3), 336–348. DOI: 10.1111/jcpp.13190. 5, 61

Ofiesh, N. S., Rice, C. J., Long, E. M., Merchant, D. C., and Gajar, A. H. (2002). Service delivery for postsecondary students with disabilities: A survey of assistive technology use across disabilities. *College Student Journal*, 36(1). 75, 79

Ogbonnaya-Ogburu, I. F., Toyama, K., and Dillahunt, T. (2018). Returning citizens' job search and technology use: Preliminary findings. In *Companion of the 2018 ACM Conference on Computer Supported Cooperative Work and Social Computing* (pp. 365–368). DOI: 10.1145/3272973.3274098. 82

Olthuis, J. V., McGrath, P. J., Cunningham, C. E., Boyle, M. H., Lingley-Pottie, P., Reid, G. J, 57., Bagnell, A., Lipman, E. L., Turner, K., Corkum, P., Steward, S. J., Berrigan, P., and Sdao-Jarvie, K. (2018). Distance-delivered parent training for childhood disruptive behavior (Strongest FamiliesTM): A randomized controlled trial and economic analysis. *Journal of Abnormal Child Psychology* 46(8): 1613–1629. DOI: 10.1007/s10802-018-0413-y. 18

Ota, K. R. and DuPaul. G. J. (2002). Task engagement and mathematics performance in children with attention deficit hyperactivity disorder: Effects of supplemental computer instruction. *School Psychology Quarterly*. 17, 242–257. DOI: 10.1521/scpq.17.3.242.20881. 69

Page, Z. E., Barrington, S., Edwards, J., and Barnett, L. M. (2017). Do active video games benefit the motor skill development of non-typically developing children and adolescents: A systematic review. *Journal of Science and Medicine in Sport*, 20(12), 1087–1100. DOI: 10.1016/j.jsams.2017.05.001. 95

Palsbo, S. E. and Hood-Szivek, P. (2012). Effect of robotic-assisted three-dimensional repetitive motion to improve hand motor function and control in children with handwriting deficits: A nonrandomized phase 2 device trial. *American Journal of Occupational Therapy*, 66(6), 682–690. DOI: 10.5014/ajot.2012.004556. 18, 90

Park, M. K. and Kim, H. (2018). Effect of interactive metronome training on postural control and hand writing performance of children with attention deficit hyperactivity disorder (ADHD): Single subject research. *The Journal of Korean Academy of Sensory Integration*, 16(1), 14–24. DOI: 10.18064/JKASI.2018.16.1.014. 89

Park, P., Shin, Y. M., Byun, H. Y., and Yang, J. (2009). Preliminary children health care service using ubiquitous technology. In *2009 Fifth International Joint Conference on INC, IMS and IDC* (pp. 1053–1057). IEEE. DOI: 10.1109/NCM.2009.275. 18

Park, S. (2019). 11 bad*ss women who are thriving with ADHD. *Popsugar.Fitness*. https://www.popsugar.com/fitness/Famous-Women-ADHD-46084806. 4

Park, Y. Y. and Choi, Y. J. (2017). Effects of interactive metronome training on timing, attention, working memory, and processing speed in children with ADHD: a case study of two children. *Journal of Physical Therapy Science*, 29(12), 2165–2167. DOI: 10.1589/jpts.29.2165. 89

Parker, D. R. and Banerjee, M. (2007). Leveling the digital playing field: Assessing the learning technology needs of college-bound students with LD and/or ADHD. *Assessment for Effective Intervention*, 33(1), 5–14. DOI: 10.1177/15345084070330010201. 71

Parker, D. R. and Boutelle, K. (2009). Executive function coaching for college students with learning disabilities and ADHD: A new approach for fostering self-determination. *Learning Disabilities Research and Practice*, 24(4), 204–215. DOI: 10.1111/j.1540-5826.2009.00294.x. 53

Parker, D. R., White, C. E., Collins, L., Banerjee, M., and McGuire, J. M. (2009b). Learning technologies management system (LiTMS): A multidimensional service delivery model for college students with learning disabilities and ADHD. *Journal of Postsecondary Education and Disability*, 22(2), 130–136. 72

Parlakian, R. (2003). *Before the ABCs: Promoting School Readiness in Infants and Toddlers*. Zero to Three, Herdon, VA. 45

Parsons, T. D., Bowerly, T., Buckwalter, J. G., and Rizzo, A. A. (2007). A controlled clinical comparison of attention performance in children with ADHD in a virtual reality classroom compared to standard *Neuropsychologica*l methods. *Child Neuropsychology*, 13(4), 363–381. DOI: 10.1080/13825580600943473. 18, 24

Paul, M. (2003). *It's Hard to Make a Difference When You Can't Find Your Keys: The Seven-Step Path to Becoming Truly Organized*. Penguin. 65

Pavri, S. (2004). General and special education teachers' preparation needs in providing social support: A needs assessment. *Teacher Education and Special Education: The Journal of the Teacher Education Division of the Council for Exceptional Children*, 27, 433–443. DOI: 10.1177/088840640402700410. 66

Pellegrini, S., Murphy, M., and Lovett, E. (2020). The QbTest for ADHD assessment: Impact and implementation in child and adolescent mental health services. *Children and Youth Services Review*. DOI: 10.1016/j.childyouth.2020.105032. 27

Peña, O., Cibrian, F. L., and Tentori, M. (2020). Circus in Motion: a multimodal exergame supporting vestibular therapy for children with autism. *Journal on Multimodal User Interfaces*, 1–17. DOI: 10.1007/s12193-020-00345-9. 94

Peng, X., Lin, P., Zhang, T., and Wang, J. (2013). Extreme learning machine-based classification of ADHD using brain structural MRI data. *PLoS ONE*, 8(11), e79476. DOI: 10.1371/journal.pone.0079476. 32

Pesce, C., Lakes, K.D. Stodden, D., and Marchetti, R. (2020). Fostering self-control development with a designed intervention in physical education: a two-year class-randomized trial. *Child Development*. DOI: 10.1111/cdev.13445. 92

Peters, R. E., Pak, R., Abowd, G. D., Fisk, A. D., and Rogers, W. A. (2004). *Finding Lost Objects: Informing the Design of Ubiquitous Computing Services for the Home*. Atlanta, GA: Georgia Institute of Technology. 78

Pina, L., Rowan, K., Roseway, A., Johns, P., Hayes, G. R., and Czerwinski, M. (2014). In situ cues for ADHD parenting strategies using mobile technology. *Proceedings - PERVASIVE-HEALTH 2014: 8th International Conference on Pervasive Computing Technologies for Healthcare* (pp. 17–24). DOI: 10.4108/icst.pervasivehealth.2014.254958. 18, 58

Pinsky, S. C. (2012). *Organizing Solutions for Oeople with ADHD: Tips and Tools to Help You Take Charge of Your Life and Get Organized*. Fair Winds Press. 65

Piper, A. M., O'Brien, E., Morris, M. R., and Winograd, T. (2006). SIDES: a cooperative tabletop computer game for social skills development. In *Proceedings of the 2006 20th Anniversary Conference on Computer Supported Cooperative Work* (pp. 1–10). DOI: 10.1145/1180875.1180877. 48

Pitts, M., Mangle, L., and Asherson, P. (2015). Impairments, diagnosis and treatments associated with attention-deficit/hyperactivity disorder (ADHD) in the U.K. adults: Results from the lifetime impairments survey. *Archive of Psychiatric Nursing*, 29, 56–63. DOI: 10.1016/j.apnu.2014.10.001. 81

Polanczyk, G. V., Willcutt, E. G., Salum, G. A., Kieling, C., and Rohde, L. A. (2014). ADHD prevalence estimates across three decades: an updated systematic review and meta-regression analysis. *International Journal of Epidemiology*, 43(2), 434–442. DOI: 10.1093/ije/dyt261. 1

Pollak, Y., Weiss, P. L., Rizzo, A. A., Weizer, M., Shriki, L., Shalev, R. S., and Gross-Tsur, V. (2009). The Utility of a Continuous Performance Test Embedded in Virtual Reality in Measuring ADHD-Related Deficits. *Journal of Developmental and Behavioral Pediatrics*, 30(1), 2–6. DOI: 10.1097/DBP.0013e3181969b22. 18, 24

Porrino, L. J., Rapoport, J. L., Behar, D., Sceery, W., Ismond, D. R., and Bunney, W. E. (1983). A naturalistic assessment of the motor activity of hyperactive boys: I. Comparison with normal controls. *Archives of General Psychiatry*, 40(6), 681–687. DOI: 10.1001/archpsyc.1983.04390010091012. 30

Powell, L., Parker, J., and Harpin, V. (2018). What is the level of evidence for the use of currently available technologies in facilitating the self-management of difficulties associated with ADHD in children and young people? A systematic review. *European Child and Adolescent Psychiatry*, 27(11), 1391–1412. DOI: 10.1007/s00787-017-1092-x. 6

Powell, L., Parker, J., and Harpin, V. (2017). ADHD: Is there an app for that? A suitability assessment of apps for the parents of children and young people with ADHD. *JMIR mHealth and uHealth*, 5(10), e149. DOI: 10.2196/mhealth.7941. 6

Prevatt, F. and Yelland, S. (2013). An empirical evaluation of ADHD coaching in college students. *Journal of Attention Disorders*, 19, 666–677. DOI: 10.1177/1087054713480036. 83

Prevatt, F., Lampropoulos, G. K., Bowles, V., and Garrett, L. (2011). The use of between session assignments in ADHD coaching with college students. *Journal of Attention Disorders*, 15, 18–27. DOI: 10.1177/1087054709356181. 83

Prins, P. J., Brink, E. T., Dovis, S., Ponsioen, A., Geurts, H. M., De Vries, M., and Van Der Oord, S. (2013). "Braingame Brian": toward an executive function training program with game elements for children with ADHD and cognitive control problems. *GAMES FOR HEALTH: Research, Development, and Clinical Applications*, 2(1), 44–49. DOI: 10.1089/g4h.2013.0004. 18

Puskar, K. R., Lamb, J., Boneysteele, G., Sereika, S., Rohay, J., and Tusaie-Mumford, K. (1996). High touch meets high tech. Distance mental health screening for rural youth using Teleform. *Computers in Nursing*, 14(6), 323–9. 5

Ra, C. K., Cho, J., Stone, M. D., De La Cerda, J., Goldenson, N. I., Moroney, E., Tung, I., Lee, S. S., and Leventhal, A. M. (2018). Association of digital media use with subsequent symptoms of attention-deficit/hyperactivity disorder among adolescents. *JAMA*, 320(3), 255–263. DOI: 10.1001/jama.2018.8931. 67

Rabiner, D. L., Murray, D. W., Skinner, A. T., and Malone, P. S. (2010). A randomized trial of two promising computer-based interventions for students with attention difficulties. *Journal of Abnormal Child Psychology*, 38(1), 131–142. DOI: 10.1007/s10802-009-9353-x. 69

Ralston, A. L., Andrews III, A. R., and Hope, D. A. (2019). Fulfilling the promise of mental health technology to reduce public health disparities: Review and research agenda. *Clinical Psychology: Science and Practice*, 26(1), e12277. DOI: 10.1111/cpsp.12277. 5

Rangarajan, B., Suresh, S., and Mahanand, B. S. (2014). Identification of potential biomarkers in the hippocampus region for the diagnosis of ADHD using PBL-McRBFN approach. In *2014 13th International Conference on Control Automation Robotics and Vision* (ICARCV) (pp. 17–22). IEEE. DOI: 10.1109/ICARCV.2014.7064272. 32

Rapport, M. D., Orban, S. A., Kofler, M. J., and Friedman, L. M. (2013). Do programs designed to train working memory, other executive functions, and attention benefit children with ADHD? A meta-analytic review of cognitive, academic, and behavioral outcomes. *Clinical Psychology Review*, 33(8), 1237–1252. DOI: 10.1016/j.cpr.2013.08.005. 39, 40

Raver, C. C. (2004). Placing emotional self-regulation in sociocultural and socioeconomic contexts. *Child Development*, 75(2), 346–353. DOI: 10.1111/j.1467-8624.2004.00676.x. 55

Reid, R. C., Bramen, J. E., Anderson, A., and Cohen, M. S. (2014). Mindfulness, emotional dysregulation, impulsivity, and stress proneness among hypersexual patients. *Journal of Clinical Psychology*, 70(4), 313–321. DOI: 10.1002/jclp.22027. 83

Reid, R., Trout, A. L., and Schartz, M. (2005). Self-regulation interventions for children with attention deficit/hyperactivity disorder. *Exceptional Children*, 71(4), 361. 55

Rey, A. (1942). L'examen psychologique dans les cas d'encephalopathie traumatique. *Archives de Psychologie*, 28, 112. 28

Riaz, A., Asad, M., Alonso, E., and Slabaugh, G. (2018). Fusion of fMRI and non-imaging data for ADHD classification. *Computerized Medical Imaging and Graphics*, 65, 115–128. DOI: 10.1016/j.compmedimag.2017.10.002. 32

Ricci, M., Terribili, M., Giannini, F., Errico, V., Pallotti, A., Galasso, C., Tomasello, L., Sias, S., and Saggio, G. (2019). Wearable-based electronics to objectively support diagnosis of motor impairments in school-aged children. *Journal of Biomechanics*, 83, 243–252. DOI: 10.1016/j.jbiomech.2018.12.005. 31

Riek, L. D. (2016). Robotics technology in mental health care. In *Artificial Intelligence in Behavioral and Mental Health Care* (pp. 185–203). Academic Press. DOI: 10.1016/B978-0-12-420248-1.00008-8. 60

Riemer-Reiss, M. L. (2000). Utilizing distance technology for mental health counseling. *Journal of Mental Health Counseling*, 22(3). 5

Rijo, R., Costa, P., Machado, P., Bastos, D., Matos, P., Silva, A., Ferrinho, J., Almeida, N., Oliveira, A., Xavier, S., Santos, S., Oliveira, C., Brites, S., Martins, V., Pereira, A. and Fernandes, S. (2015). Mysterious bones unearthed: Development of an online therapeutic serious game for children with attention deficit-hyperactivity disorder. In *Procedia Computer Science* (Vol. 64, pp. 1208–1216). Elsevier Masson SAS. DOI: 10.1016/j.procs.2015.08.512. 18

Ripat, J. D. and Woodgate, R. L. (2017). The importance of assistive technology in the productivity pursuits of young adults with disabilities. *Work*, 57(4), 455–468. DOI: 10.3233/WOR-172580. 85

Rizzo, A. A. and Buckwalter, J. G. (1997). Virtual reality and cognitive assessment and rehabilitation: The state of the art. *Studies in Health Technology and Informatics*, 44, 123–145. 24

Rizzo, A., Bowerly, T., Buckwalter, J., Klimchuk, D., Mitura, R., and Parsons, T. (2006). A virtual reality scenario for all seasons: The virtual classroom. *CNS Spectrums*, 11, 35–44. DOI: 10.1017/S1092852900024196. 24

Rizzo, A., Bowerly, T., Buckwalter, J., Shahabi, C., Kim, L., Sharifzadeh, M., et al. (2003). Virtual environments for the assessment of cognitive and functional abilities: Results from the virtual classroom and office. *The Clinical Neuropsychologist*, 17, 100–100. DOI: 10.1037/e351772004-001. 24

Rizzo, A., Bowerly, T., Shahabi, C., Buckwalter, J., Klimchuk, D., and Mitura, R. (2004). Diagnosing attention disorders in a virtual classroom. *Computer*, 37, 87–89. DOI: 10.1109/MC.2004.23. 24

Rizzo, A., Buckwalter, J., Bowerly, T., van der Zaag, C., Humphrey, L., Neumann, U., et al. (2000). The virtual classroom: A virtual reality environment for the assessment and rehabilitation of attention deficits. *Cyberpsychology and Behavior*, 3, 483–499. DOI: 10.1089/10949310050078940. 24

Rodríguez, C., Areces, D., García, T., Cueli, M., and González-Castro, P. (2018). Comparison between two continuous performance tests for identifying ADHD: Traditional vs. Virtual reality. *International Journal of Clinical and Health Psychology : (IJCHP)*, 18(3), 254–263. DOI: 10.1016/j.ijchp.2018.06.003. 18, 23, 25

Roessler, R. T., Rumrill Jr, P. D., Rumrill, S. P., Minton, D. L., Hendricks, D. J., Sampson, E., Stauffer, C., Scherer, M. J., Nardone, A., Leopold, A., Jacobs, K., and Elias, E. (2017). Qualitative case studies of professional-level workers with traumatic brain injuries: A contextual approach to job accommodation and retention. *Work*, 58(1), 3–14. DOI: 10.3233/WOR-162601. 84

Rosario-Hernández, E., Rovira-Millán, L., Santiago-Pacheco, E., Arzola-Berrios, X., Padovani, C. M., Francesquini-Oquendo, S., Soto-Franceschini, J. A., Pons-Madera, J. I., Peña, L., and Vélez, E. (2020). ADHD and its effects on job performance: A moderated mediation model. *Revista Caribeña de Psicología*, 4(1), 1–25. DOI: 10.37226/rcp.2020/01. 82

Rose, D. H., Harbour, W. S., Johnston, C. S., Daley, S. G., and Abarbanell, L. (2006). Universal design for learning in postsecondary education: Reflections on principles and their application. *Journal of Postsecondary Education and Disability*, 19(2), 135–151. 72

Rosen, L. D. and Weil, M. M. (1997). *The Mental Health Technology Bible with Cdrom*. John Wiley and Sons, Inc. 5

Rosen, P. J. and Factor, P. I. (2015). Emotional impulsivity and emotional and behavioral difficulties among children with ADHD: An ecological momentary assessment study. *Journal of Attention Disorders*, 19(9), 779–793. DOI: 10.1177/1087054712463064. 52

Rosen, P. J., Epstein, J. N., and Van Orden, G. (2013). I know it when I quantify it: Ecological momentary assessment and recurrence quantification analysis of emotion dysregulation in children with ADHD. *ADHD Attention Deficit and Hyperactivity Disorders*, 5(3), 283–294. DOI: 10.1007/s12402-013-0101-2. 67

Rosenblatt, J. (2019). More screen time linked to higher risk of ADHD in preschool-aged children. *ABC News.* https://abcnews.go.com/Health/screen-time-linked-higher-risk-adhd-preschool-aged/story?id=62429157. 5

Rossiter, D. T. R. and La Vaque, T. J. (1995). A comparison of EEG biofeedback and psychostimulants in treating attention deficit/hyperactivity disorders. *Journal of Neurotherapy*, 1(1), 48–59. DOI: 10.1300/J184v01n01_07. 63

Rothbaum, B. O., Meadows, E. A., Resick, P., and Foy, D. W. (2000). Cognitive-behavioral therapy. In E. B. Foa, T. M. Keane, and M. J. Friedman (Eds.), *Effective Treatments for PTSD: Practice Guidelines from the International Society for Traumatic Stress Studies* (pp. 320–325). Guilford Press. 46

Ruiz-Manrique, G., Tajima-Pozo, K., and Montañes-Rada, F. (2014). Case Report: ADHD Trainer: the mobile application that enhances cognitive skills in ADHD patients. F1000Research, 3. DOI: 10.12688/f1000research.5689.1. 61, 67

Rusk, M. (2016). Lumosity to pay $2 million to settle FTC deceptive advertising charges for its "brain training" program. Federal Trade Commission. https://tinyurl.com/y3hxw3ej. 38

Ryan, G. S., Haroon, M., and Melvin, G. (2015). Evaluation of an educational website for parents of children with ADHD. *International Journal of Medical Informatics*, 84(11), 974–981. DOI: 10.1016/j.ijmedinf.2015.07.008. 18, 57

Sachnev, V. (2015). An efficient classification scheme for ADHD problem based on Binary Coded Genetic Algorithm and McFIS. In *2015 International Conference on Cognitive Computing and Information Processing* (CCIP) (pp. 1–6). IEEE. DOI: 10.1109/CCIP.2015.7100690. 32

Santos, F. E. G., Bastos, A. P. Z., Andrade, L. C. V., Revoredo, K., and Mattos, P. (2011). Assessment of ADHD through a computer game: An experiment with a sample of students. *Proceedings - 2011 3rd International Conferenceon Games and Virtual Worlds for Serious Applications, VS-Games 2011* (pp. 104–111). DOI: 10.1109/VS-GAMES.2011.21. 18, 26

Sarafpour, M., Shirazi, S. Y., Shirazi, E., Ghazaei, F., and Parnianpour, Z. (2018). Postural balance performance of children with ADHD, with and without medication: a quantitative approach. In *2018 40th Annual International Conference of the IEEE Engineering in Medicine and Biology Society* (EMBC) (pp. 2100–2103). IEEE. DOI: 10.1109/EMBC.2018.8512636. 34

Schaeffer, B., Thomas, J. C., and Hersen, M. (2004). Learning disabilities and attention deficits in the workplace. *Psychopathology in the Workplace: Recognition and Adaptation*. New York: Brunner-Routledge (pp. 201–224). 81

Schafer, E. C., Mathewsa, L., Mehtab, S., Hilla, M., Munoza, A., Bishopa, R., and Moloney, M. (2013). Personal FM systems for children with autism spectrum disorders (ASD) and/ or attention-deficit hyperactivity disorder (ADHD): An initial investigation. *Journal of Communication Disorders*, 46(1), 30–52. DOI: 10.1016/j.jcomdis.2012.09.002. 82

Schnieders, C. A., Gerber, P. J., and Goldberg, R. J. (2015). Integrating findings of studies of successful adults with learning disabilities: A new comprehensive model for researchers and practitioners. *Journal of Career Planning and Adult Development*, 31(4). 82

Schuck, S., Emmerson, N., Ziv, H., Collins, P., Arastoo, S., Warschauer, M., Crinella, F., and Lakes, K. (2016). Designing an iPad app to monitor and improve classroom behavior for children with ADHD: ISelfControl feasibility and pilot studies. *PLoS ONE*, 11(10), 1–13. DOI: 10.1371/journal.pone.0164229. 19, 59

Schwartz, M. S. and Andrasik, F. (Eds.). (2017). *Biofeedback: A Practitioner's Guide*. Guilford Publications. 62

Scott, S. S., McGuire, J. M., and Shaw, S. (2001). *Principles of Universal Design for Instruction*. Storrs, CT: University of Connecticut, Center on Postsecondary Education and Disability. 72

Sehlin, H., Hedman Ahlström, B., Andersson, G., and Wentz, E. (2018). Experiences of an internet-based support and coaching model for adolescents and young adults with ADHD and autism spectrum disorder -a qualitative study. *BMC Psychiatry*, 18(1), 1–13. DOI: 10.1186/s12888-018-1599-9. 19, 53

Seli, P., Smallwood, J., Cheyne, J. A., and Smilek, D. (2014). On the relation of mind wandering and ADHD symptomatology. *Psychology Bulletin Review*, 22, 629–636. DOI: 10.375/s13423-014-0793-0. 82

Selk, J., Bartow, T., and Rudy, M. (2015). *Organize Tomorrow Today: 8 Ways to Retrain Your Mind to Optimize Performance at Work and in Life*. Da Capo Lifelong Books. 65

Sen, B., Borle, N. C., Greiner, R., and Brown, M. R. (2018). A general prediction model for the detection of ADHD and Autism using structural and functional MRI. *PLoS ONE*, 13(4), e0194856. DOI: 10.1371/journal.pone.0194856. 32

Seymour, K. E., Chronis-Tuscano, A., Halldorsdottir, T., Stupica, B., Owens, K., and Sacks, T. (2012). Emotion regulation mediates the relationship between ADHD and depressive symptoms in youth. *Journal of Abnormal Child Psychology*, 40, 595–606. DOI: 10.1007/s10802-011-9593-4. 52

Shaffer, R. J., Jacokes, L. E., Cassily, J. F., Greenspan, S. I., Tuchman, R. F., and Stemmer, P. J. (2001). Effect of Interactive Metronome® training on children with ADHD. *American Journal of Occupational Therapy*, 55(2), 155–162. DOI: 10.5014/ajot.55.2.155. 89

Shahin, S., Reitzel, M., Di Rezze, B., Ahmed, S., and Anaby, D. (2020). Environmental factors that impact the workplace participation of transition-aged young adults with brain-based disabilities: A scoping review. *International Journal of Environmental Research and Public Health*, 17(7), 2378. DOI: 10.3390/ijerph17072378. 81

Sharma, S., Varkey, B., Achary, K., Hakulinen, J., Turunen, M., Heimonen, T., Svivasta, S., and Rajput, N. (2018). Designing gesture-based applications for individuals with developmental disabilities: Guidelines from user studies in India. *ACM Transactions on Accessible Computing*, 11(1), 1–27. DOI: 10.1145/3161710. 91

Shema-Shiratzky, S., Brozgol, M., Cornejo-Thumm, P., Geva-Dayan, K., Rotstein, M., Leitner, Y., Hausdorff, J. M., and Mirelman, A. (2019). Virtual reality training to enhance behavior and cognitive function among children with attention-deficit/hyperactivity disorder: brief report. *Developmental Neurorehabilitation*, 22(6), 431–436. DOI: 10.1080/17518423.2018.1476602. 19, 42, 91

Shiffman, S., Stone, A. A., and Hufford, M. R. (2008). Ecological momentary assessment. *Annual Review of Clinical Psychology*, 4, 1–32. DOI: 10.1146/annurev.clinpsy.3.022806.091415. 23, 52

Shih, C. H., Wang, S. H., and Wang, Y. T. (2014). Assisting children with Attention Deficit Hyperactivity Disorder to reduce the hyperactive behavior of arbitrary standing in class with a Nintendo Wii remote controller through an active reminder and preferred reward stimulation. *Research in Developmental Disabilities*, 35(9), 2069–2076. DOI: 10.1016/j.ridd.2014.05.007. 19, 60, 90

Shih, C. H., Yeh, J. C., Shih, C. T., and Chang, M. L. (2011). Assisting children with attention deficit hyperactivity disorder actively reduces limb hyperactive behavior with a Nintendo Wii remote controller through controlling environmental stimulation. *Research in Developmental Disabilities*. DOI: 10.1016/j.ridd.2011.02.014. 19, 60, 90

Shih, C.-H. (2011). Assisting people with attention deficit hyperactivity disorder by actively reducing limb hyperactive behavior with a gyration air mouse through a controlled environmental stimulation. *Research in Developmental Disabilities*, 32(1), 30–36 DOI: 10.1016/j.ridd.2010.08.009. 19, 60, 90

Shrieber, B. and Seifert, T. (2009). College students with learning disabilities and/or ADHD use of a handheld computer compared to conventional planners. In *Proceedings of the Chairs*

Conference on Instructional Technologies Research. Learning in the Technological Era, Y. Eshet-Alkalai, A. Caspi, S. Eden, N. Geri, Y. Yair (Eds.), Raanana: The Open University of Israel. 70

Shuren, J., Patel, B., and Gottlieb, S. (2018). FDA regulation of mobile medical apps. *JAMA*, 320(4), 337–338. DOI: 10.1001/jama.2018.8832. 99

Silva, A. P. and Frère, A. F. (2011). Virtual environment to quantify the influence of colour stimuli on the performance of tasks requiring attention. *Biomedical Engineering Online*, 10, 74. DOI: 10.1186/1475-925X-10-74. 24

Slovák, P. and Fitzpatrick, G. (2015). Teaching and developing social and emotional skills with technology. *ACM Transactions on Computer-Human Interaction* (TOCHI), 22(4), 1–34. DOI: 10.1145/2744195. 46

Slovák, P., Gilad-Bachrach, R., and Fitzpatrick, G. (2015). Designing social and emotional skills training: The challenges and opportunities for technology support. In *Proceedings of the 33rd Annual ACM Conference on Human Factors in Computing Systems* (pp. 2797–2800). DOI: 10.1145/2702123.2702385. 46

Slovák, P., Rowan, K., Frauenberger, C., Gilad-Bachrach, R., Doces, M., Smith, B., Kamb, R., and Fitzpatrick, G. (2016). Scaffolding the scaffolding: Supporting children's social-emotional learning at home. In *Proceedings of the 19th ACM Conference on Computer-Supported Cooperative Work and Social Computing* (pp. 1751–1765). DOI: 10.1145/2818048.2820007. 46

Smith, S. R., Chang, J., Schnoebelen, K. J., Edwards, J. W., Servesko, A. M., and Walker, S. J. (2007). Psychometrics of a simple method for scoring organizational approach on the Rey-Osterrieth complex figure. *Journal of Neuropsychology*, 1(1), 39–51. DOI: 10.1348/174866407X180800. 28

Söderqvist, H., Kajsa, E., Ahlström, B. H., and Wentz, E. (2017). The caregivers' perspectives of burden before and after an internet-based intervention of young persons with ADHD or autism spectrum disorder. *Scandinavian Journal of Occupational Therapy*, 24(5), 383–392. DOI: 10.1080/11038128.2016.1267258. 19, 53

Solstad Vedeler, J. and Schreuer, N. (2011). Policy in action: Stories on the 79 workplace accommodation process. *Journal of Disability Policy Studies*, 22(2), 95–105. DOI: 10.1177/1044207310395942. 84, 85

Sonne, T. and Jensen, M. M. (2016d). ChillFish: A respiration game for children with ADHD. *TEI 2016 - Proceedings of the 10th Anniversary Conference on Tangible Embedded and Embodied Interaction* (pp. 271–278). DOI: 10.1145/2839462.2839480. 19, 62

Sonne, T., Marshall, P., Müller, J., Obel, C., and Grønbæk, K. (2016a). A follow-up study of a successful assistive technology for children with ADHD and their families. *Proceedings of IDC 2016 - The 15th International Conference on Interaction Design and Children* (pp. 400–407). DOI: 10.1145/2930674.2930704. 19, 80

Sonne, T., Marshall, P., Obel, C., Thomsen, P. H., and Grønbæk, K. (2016b). An assistive technology design framework for ADHD. In *Proceedings of the 28th Australian Conference on Computer-Human Interaction* (pp. 60–70). DOI: 10.1145/3010915.3010925. 79

Sonne, T., Merritt, T., Marshall, P., Lomholt, J. J., Müller, J., and Grønbæk, K. (2017). Calming children when drawing blood using breath-based biofeedback. In *Proceedings of the 2017 Conference on Designing Interactive Systems* (pp. 725–737). DOI: 10.1145/3064663.3064742. 19, 62

Sonne, T., Müller, J., Marshall, P., Obel, C., and Grønbæk, K. (2016c). Changing family practices with assistive technology: MOBERO improves morning and bedtime routines for children with ADHD. In *Proceedings of the 2016 CHI Conference on Human Factors in Computing Systems* (pp. 152–164). DOI: 10.1145/2858036.2858157. 19, 80

Sonne, T., Obel, C., and Grønbæk, K. (2015). Designing real time assistive technologies: a study of children with ADHD. In *Proceedings of the Annual Meeting of the Australian Special Interest Group for Computer Human Interaction* (pp. 34–38). DOI: 10.1145/2838739.2838815. 19, 43

Sonuga-Barke, E. J., Brandeis, D., Cortese, S., Daley, D., Ferrin, M., Holtmann, M., Stevenson, J., Danckaerts, M., van der Oord, S., Döpfner, M., Dittmann, R. W., Simonoff, E, Zuddas, A., Banaschewski, T., Buitelaar, J., Coghill, D., Hollis, C., Konofal, E., Lecendrewux, M., Wong, I. C. K., Sergeant, J., and European ADHD Guidelines Group. (2013). Nonpharmacological interventions for ADHD: systematic review and meta-analyses of randomized controlled trials of dietary and psychological treatments. *American Journal of Psychiatry*, 170(3), 275–289. DOI: 10.1176/appi.ajp.2012.12070991. 63

Spitale, M., Gelsomini, M., Beccaluva, E., Viola, L., and Garzotto, F. (2019). Meeting the needs of people with neuro-developmental disorder through a phygital approach. In *CHItaly 2019* (pp. 1–10). DOI: 10.1145/3351995.3352055. 19, 68

Steffert, B, and Steffert, T. (2010). Neurofeedback and ADHD. *ADHD in Practice*. 2:16–9. DOI: 10.1521/adhd.2011.19.2.16. 63

Stein, K. and Milne, R. (1999). Mental health technology assessment: practice based research to support evidence-based practice. *Evidence-Based Mental Health*, 2(2), 37–39. DOI: 10.1136/ebmh.2.2.37. 5

Steinau, S. and Kandemir, H. (2013). Diagnostic criteria in attention deficit hyperactivity disorders changes in DSM 5. *Frontiers in Psychiatry*, 4, 1–2. DOI: 10.3389/fpsyt.2013.00049. 82

Steinberg, E. A. and Drabick, D. A. G. (2015). A developmental psychopathology per- spective on ADHD and comorbid conditions: The role of emotion regulation. *Child Psychiatry and Human Development*, 46, 951–966. DOI: 10.1007/s10578-015-0534-2. 52

Steiner, N. J., Frenette, E. C., Rene, K. M., Brennan, R. T., and Perrin, E. C. (2014a). In-school neurofeedback training for ADHD: Sustained improvements from a randomized control trial. *Pediatrics*, 133(3), 483–492. DOI: 10.1542/peds.2013-2059. 41, 62

Steiner, N. J., Frenette, E. C., Rene, K. M., Brennan, R. T., and Perrin, E. C. (2014b). Neurofeedback and cognitive attention training for children with ADHD in schools. *Journal of Developmental and Behavioral Pediatrics*, 35(1), 18–27. DOI: 10.1097/DBP.0000000000000009. 41, 62

Stiglic, N. and Viner, R. M. (2019). Effects of screentime on the health and well-being of children and adolescents: a systematic review of reviews. *BMJ Open*, 9(1). DOI: 10.1136/bmjopen-2018-023191. 5, 61

Storebø, O. J., Andersen, M. E., Skoog, M., Hansen, S. J., Simonsen, E., Pedersen, N., Tendal, B., Callesen, J., Faltinsen, E., and Gluud, C. (2019). Social skills training for attention deficit hyperactivity disorder (ADHD) in children aged 5 to 18 years. *Cochrane Database of Systematic Reviews*, (6). DOI: 10.1002/14651858.CD008223.pub3. 46

Strine, T. W., Lesesne, C. A., Okoro, C. A., McGuire, L. C., Chapman, D. P., Balluz, L. S., and Mokdad, A. H. (2006). Peer reviewed: emotional and behavioral difficulties and impairments in everyday functioning among children with a history of attention-deficit/hyperactivity disorder. *Preventing Chronic Disease*, 3(2). 46

Surman, C. B., Biederman, J., Spencer, T., Miller, C. A., McDermott, K. M., and Faraone, S. V. (2013). Understanding deficient emotional self-regulation in adults with attention deficit hyperactivity disorder: A controlled study. *ADHD Attention-Deficit Hyperactivity Disorder*, 5, 273–381. DOI: 10.1007/s12402-012-0100-8. 82

Swanson, J. M., Wigal, T. L., and Lakes, K. D. (2009). DSM V and the future of diagnosis of attention deficit hyperactivity disorder. *Current Psychiatry Reports*, 11(5), 399–406. DOI: 10.1007/s11920-009-0060-7. 3

Swauger, S. (2020). Software that monitors students during tests perpetuates inequality and violates their privacy. *MIT Technology Review*. https://www.technologyreview.com/2020/08/07/1006132/software-algorithms-proctoring-online-tests-ai-ethics/. 73

Swift, C. and Levin, G. (1987). Empowerment: An emerging mental health technology. *Journal of Primary Prevention*, 8(1-2), 71–94. DOI: 10.1007/BF01695019. 6

Tam, V., Gelsomini, M., and Garzotto, F. (2017). Polipo - A tangible toy for children with neurodevelopmental disorders. *TEI 2017 - Proceedings of the 11th International Conference on Tangible, Embedded, and Embodied Interaction* (pp. 11–20). DOI: 10.1145/3024969.3025006. 19, 90

Tamana, S. K., Ezeugwu, V., Chikuma, J., Lefebvre, D. L., Azad, M. B., Moraes, T. J., Subbarao, P., Becker, A., B., Turvey, S. E., Sears, M. R., Dick, B. D., Carson, V., Rasmussen, C., CHILD Study Investigators, Pei, J., and Mandhane, P. J. (2019). Screen-time is associated with inattention problems in preschoolers: Results from the CHILD birth cohort study. *PLoS ONE*, 14(4), e0213995. DOI: 10.1371/journal.pone.0213995. 5

Tamura, K., Kato, S., Yamakita, T., Kimura, K., and Itoh, H. (2003). A system of dialogical mental health care with sensibility technology communication robot. In MHS2003. *Proceedings of 2003 International Symposium on Micromechatronics and Human Science* (IEEE Cat. No. 03TH8717) (pp. 67–71). IEEE. DOI: 10.1109/MHS.2003.1249911. 60

Tan, L., Guo, X., Ren, S., Epstein, J. N., and Lu, L. J. (2017). A computational model for the automatic diagnosis of attention deficit hyperactivity disorder based on functional brain volume. *Frontiers in Computational Neuroscience*, 11, 75. DOI: 10.3389/fncom.2017.00075. 32

Tan, Y., Zhu, D., Gao, H., Lin, T. W., Wu, H. K., Yeh, S. C., and Hsu, T. Y. (2019). Virtual classroom: An ADHD assessment and diagnosis system based on virtual reality. In *Proceedings - 2019 IEEE International Conference on Industrial Cyber Physical Systems, ICPS 2019* (pp. 203–208). DOI: 10.1109/ICPHYS.2019.8780300. 19, 24

Tanaka, H., Sakriani, S., Neubig, G., Toda, T., Negoro, H., Iwasaka, H., and Nakamura, S. (2016). Teaching social communication skills through human-agent interaction. *ACM Transactions on Interactive Intelligent Systems* (TiiS), 6(2), 1–26. DOI: 10.1145/2937757. 46

Tenev, A., Markovska-Simoska, S., Kocarev, L., Pop-Jordanov, J., Müller, A., and Candrian, G. (2014). Machine learning approach for classification of ADHD adults. *International Journal of Psychophysiology*, 93(1), 162–166. DOI: 10.1016/j.ijpsycho.2013.01.008. 33

Thomas, L., Briggs, P., and Little, L. (2010). A case study: The impact of using location-based services with a behaviour-disordered child. *NordiCHI 2010: Extending Boundaries - Proceedings of the 6th Nordic Conference on Human-Computer Interaction* (pp. 503–510). DOI: 10.1145/1868914.1868971. 19, 79

Thomas, R., Sanders, S., Doust, J., Beller, E., and Glasziou, P. (2015). Prevalence of attention-deficit/hyperactivity disorder: a systematic review and meta-analysis. *Pediatrics*, 135(4), e994–e1001. DOI: 10.1542/peds.2014-3482. 1

Ting, V. and Weiss, J. A. (2017). Emotion regulation and parent co-regulation in children with autism spectrum disorder. *Journal of Autism and Developmental Disorders*, 47(3), 680–689. DOI: 10.1007/s10803-016-3009-9. 56

Tobar-Munoz, H., Fabregat, R., and Baldiris, S. (2014). Using a videogame with augmented reality for an inclusive logical skills learning session. *2014 International Symposium on Computers in Education, SIIE 2014*, 189–194. DOI: 10.1109/SIIE.2014.7017728. 70

Toldrá, R. C. and Santos, M. C. (2013). People with disabilities in the labor market: Facilitators and barriers. *Work*, 45(4), 553–563. DOI: 10.3233/WOR-131641. 82

Torous, J., Nicholas, J., Larsen, M. E., Firth, J., and Christensen, H. (2018). Clinical review of user engagement with mental health smartphone apps: evidence, theory and improvements. *Evidence-Based Mental Health*, 21(3), 116–119. DOI: 10.1136/eb-2018-102891. 5

Toshniwal, S., Dey, P., Rajput, N., and Srivastava, S. (2015). VibRein: An engaging and assistive mobile learning companion for students with intellectual disabilities. *OzCHI 2015: Being Human - Conference Proceedings* (pp. 20–28). DOI: 10.1145/2838739.2838751. 19, 72

Tucha, O. (2017). Supporting patients with ADHD: Missed opportunities? *Attention Deficit Hyperactivity Disorder*, 9, 69–71. DOI: 10.007/s12402-017-0233-x. 83

Tuttle, L. J., Ahmann, E., and Wright, S. P. (2016). Emerging evidence for the efficacy of ADHD coaching. Poster presented at the annual meeting of the American Professional Society of ADHD and Related Disorders, Washington, DC. 83

Twenge, J. M. (2017). Have smartphones destroyed a generation? *The Atlantic*. theatlantic.com/magazine/archive/2017/09/has-the-smartphone-destroyed-a-generation/534198/. 5

Twenge, J. M., Joiner, T. E., Rogers, M. L., and Martin, G. N. (2018). Increases in depressive symptoms, suicide-related outcomes, and suicide rates among US adolescents after 2010 and links to increased new media screen time. *Clinical Psychological Science*, 6(1), 3–17. DOI: 10.1177/2167702617723376. 5

Twenge, J. M. and Campbell, K. W. (2018). Associations between screen time and lower psychological well-being among children and adolescents: Evidence from a population-based study. *Preventive Medicine Reports*, 12, 271 DOI: 10.1016/j.pmedr.2018.10.003. 5

Uekermann, J., Kraemer, M., Abdel-Hamid, M., Schimmelmann, B. G., Hebebrand, J., Daum, I., Wiltfang, J., and Kis, B. (2010). Social cognition in attention-deficit hyperactivity disorder (ADHD). *Neuroscience and Biobehavioral Reviews*, 34(5), 734–743. DOI: 10.1016/j. neubiorev.2009.10.009. 45

Ulberstad, F. (2016). *QbTest Technical Manual*. Stockholm, Sweden: Qbtech AB. 23

Ulberstad, F., Boström, H., Chavanon, M. L., Knollmann, M., Wiley, J., Christiansen, H., and Thorell, L. B. (2020). Objective measurement of attention deficit hyperactivity disorder symptoms outside the clinic using the QbCheck: Reliability and validity. *International Journal of Methods in Psychiatric Research*, e1822. DOI: 10.1002/mpr.1822. 23

Ulgado, R. R., Nguyen, K., Custodio, V. E., Waterhouse, A., Weiner, R., and Hayes, G. (2013). VidCoach: a mobile video modeling system for youth with special needs. In *Proceedings of the 12th International Conference on Interaction Design and Children* (pp. 581–584). DOI: 10.1145/2485760.2485870. 82

UNESCO (2017). *A Guide for Ensuring Inclusion and Equity in Education*, UNESCO, also available in French, Spanish and Chinese Retrieved from https://en.unesco.org/themes/inclusion-in-education/disabilities/resources. 65

United Nations. Article 24-Education. (2006). Retrieved from: https://www.un.org/development/desa/disabilities/convention-on-the-rights-of-persons-with-disabilities/article-24-education.html. 65

Vaidyam, A. N., Wisniewski, H., Halamka, J. D., Kashavan, M. S., and Torous, J. B. (2019). Chatbots and conversational agents in mental health: a review of the psychiatric landscape. *The Canadian Journal of Psychiatry*, 64(7), 456–464. DOI: 10.1177/0706743719828977. 60

Van Doren, J., Arns, M., Heinrich, H., Vollebregt, M. A., Strehl, U., and Loo, S. K. (2019). Sustained effects of neurofeedback in ADHD: a systematic review and meta-analysis. *European Child and Adolescent Psychiatry*, 28(3), 293–305. DOI: 10.1007/s00787-018-1121-4. 41

Van Rooij, M., Lobel, A., Harris, O., Smit, N., and Granic, I. (2016). DEEP: A biofeedback virtual reality game for children at-risk for anxiety. In *Proceedings of the 2016 CHI Conference Extended Abstracts on Human Factors in Computing Systems* (pp. 1989–1997). DOI: 10.1145/2851581.2892452. 62

Vazou, S., Pesce, C., Lakes, K., and Smiley-Oyen, A. (2019). More than one road leads to Rome: a narrative review and meta-analysis of physical activity intervention effects on cognition in youth. *International Journal of Sport and Exercise Psychology*, 17(2), 153–178. DOI: 10.1080/1612197X.2016.1223423. 91, 92, 93

Verheul, I., Rietdijk, W., Block, J., Franken, H. L., and Thurik, R. (2016). The association between attention-deficit/hyperactivity (ADHD) symptoms and self-employment. *Europe Journal Epidemiology*, 31, 793–801. DOI: 10.1007/s10654-016-0159-1. 82

Verret, C., Guay, M. C., Berthiaume, C., Gardiner, P., and Béliveau, L. (2012). A physical activity program improves behavior and cognitive functions in children with ADHD: an exploratory study. *Journal of Attention Disorders*, 16(1), 71–80. DOI: 10.1177/1087054710379735. 92

Visser, S. N., Danielson, M. L., Bitsko, R. H., Holbrook, J. R., Kogan, M. D., Ghandour, R. M, Perou, P. H., and Blumberg, S. J. (2014). Trends in the parent-report of health care provider-diagnosed and medicated attention-deficit/hyperactivity disorder: United States, 2003–2011. *Journal of the American Academy of Child and Adolescent Psychiatry*, 53(1), 34–46. DOI: 10.1016/j.jaac.2013.09.001. 5

Voytecki, K., Anderson, P., Semon, S., and Seok, S. (2009). Assistive technology supports for postsecondary students with disabilities. In *Society for Information Technology and Teacher Education International Conference* (pp. 3990–3995). Association for the Advancement of Computing in Education (AACE). 65

Wampold, B. E. and Imel, Z. E. (2015). *The Great Psychotherapy Debate: The Evidence for What Makes Psychotherapy Work*. Routledge. DOI: 10.4324/9780203582015. 100

Wang, Z., Sun, Y., Shen, Q., and Cao, L. (2019). Dilated 3D convolutional neural networks for brain MRI data classification. *IEEE Access*, 7, 134388–134398. DOI: 10.1109/ACCESS.2019.2941912. 32

Waring, M. E. and Lapane, K. L. (2008). Overweight in children and adolescents in relation to attention-deficit/hyperactivity disorder: Results from a national sample, *Pediatrics*, 122(1), e1–6. DOI: 10.1542/peds.2007-1955. 92

Webb, K. W., Patterson, K. B., Syverud, S. M., and Seabrooks-Blackmore, J. J. (2008). Evidenced based practices that promote transition to postsecondary education: Listening to a decade of expert voices. *Exceptionality*, 16(4), 192–206. DOI: 10.1080/09362830802412182. 70

Weerdmeester, J., Cima, M., Granic, I., Hashemian, Y., and Gotsis, M. (2016). A feasibility study on the effectiveness of a full-body videogame intervention for decreasing attention deficit hyperactivity disorder symptoms. *Games for Health Journal*, 5(4), 258–269. DOI: 10.1089/g4h.2015.0103. 42, 61, 94

Weisman, O., Schonherz, Y., Harel, T., Efron, M., Elazar, M., and Gothelf, D. (2018). Testing the efficacy of a smartphone application in improving medication adherence, among children with ADHD. *Israel Journal of Psychiatry*, 55(2), 59–64. DOI: 10.1016/s0924-977x(17)31929-6. 15, 80

Weissberg, R. P., Durlak, J. A., Domitrovich, C. E., and Gullotta, T. P. (2015). Social and Emotional Learning: Past, Present, and Future. In Durlak, J. A., Domitrovich, C. E., Weissberg, R. P., and Gullotta, T. P. (Eds.), *Handbook of Social and Emotional Learning: Research and Practice* (pp. 3–19). The Guilford Press. 46

Wen, C. K. F., Schneider, S., Stone, A. A., and Spruijt-Metz, D. (2017). Compliance with mobile ecological momentary assessment protocols in children and adolescents: a systematic review and meta-analysis. *Journal of Medical Internet Research*, 19(4), e132. DOI: 10.2196/jmir.6641. 52

Wentz, E., Nydén, A., and Krevers, B. (2012). Development of an internet-based support and coaching model for adolescents and young adults with ADHD and autism spectrum disorders: a pilot study. *European Child and Adolescent Psychiatry*, 21(11), 611–622. DOI: 10.1007/s00787-012-0297-2. 19, 53

Wiklund, J., Patzelt, H., and Dimov, D. (2016). Entrepreneurship and psychological disorders: How ADHD can be productively harnessed. *Journal of Business Venturing Insights*, 6, 14–20. DOI: 10.1016/j.jbvi.2016.07.001. 4

Wills, H. P. and Mason, B. A. (2014). Implementation of a self-monitoring application to improve on-task behavior: A high-school pilot study. *Journal of Behavioral Education*, 23(4), 421–434. DOI: 10.1007/s10864-014-9204-x. 19, 58

Wimberly, L., Reed, N., and Morris, M. (2004). Postsecondary students with learning disabilities: Barriers to accessing education-based information technology. *Information Technology and Disabilities E-Journal*, 10(1), 1–40. 72

Wood, A. C., Asherson, P., Rijsdijk, F., and Kuntsi, J. (2009). Is overactivity a core feature in ADHD? Familial and receiver operating characteristic curve analysis of mechanically assessed activity level. *Journal of the American Academy of Child and Adolescent Psychiatry*, 48(10), 1023–1030. DOI: 10.1097/CHI.0b013e3181b54612. 19, 30

Yates, T., Ostrosky, M. M., Cheatham, G. A., Fettig, A., Shaffer, L., and Santos, R. M. (2008). *Research Synthesis on Screening and Assessing Social-Emotional Competence.* The Center on the Social and Emotional Foundations for Early Learning. 45

Yeh, S. C., Tsai, C. F., Fan, Y. C., Liu, P. C., and Rizzo, A. (2012). An innovative ADHD assessment system using virtual reality. In *2012 IEEE-EMBS Conference on Biomedical Engineering and Sciences* (pp. 78–83). IEEE. DOI: 10.1109/IECBES.2012.6498026. 19, 24

Young, S., Emilsson, B., Sigurdsson, J. F., Khondoker, M., Philipp-Wiegmann, F., Baldursson, G., Olafsdottir, J., and Gudjonsson, G. (2017). A randomized controlled trial reporting functional outcomes of cognitive-behavioral therapy in medication-treated adults with

ADHD and co-morbid psychopathology. *European Archives of Psychiatry and Clinical Neuroscience*, 2, 267–276. DOI: 10.1007/s00406-016-0735-0. 82

Young, Z., Craven, M. P., Groom, M., and Crowe, J. (2014). Snappy App: a mobile continuous performance test with physical activity measurement for assessing attention deficit hyperactivity disorder. In *International Conference on Human-Computer Interaction* (pp. 363–373). Springer, Cham. DOI: 10.1007/978-3-319-07227-2_35. 25

Zhao, X., Page, T. F., Altszuler, A. R. (2019). Family burden of raising a child with ADHD. *Journal of Abnormal Child Psychology*, 47, 1327–1338. DOI: 10.1007/s10802-019-00518-5. viii

Zhou, Q., Chen, S. H., and Main, A. (2012). Commonalities and differences in the research on children's effortful control and executive function: A call for an integrated model of self-regulation. *Child Development Perspectives*, 6, 112–121. DOI: 10.1111/j.1750-8606.2011.00176.x. 37

Zimmerman, B. J. and Martinez-Pons, M. (1986). Developing a structured interview for assessing student use of self-regulated learning strategies. *American Educational Research Journal*, 23, 614–628. DOI: 10.3102/00028312023004614. 72

Zimmerman, B. J., Bonner, S., and Kovach, R. (1996). *Developing Self-Regulated Learners; Beyond Achievement to Self-Efficacy.* Washington, DC: American Psychological Association. DOI: 10.1037/10213-000. 72

Zou, L., Zheng, J., Miao, C., Mckeown, M. J., and Wang, Z. J. (2017). 3D CNN based automatic diagnosis of attention deficit hyperactivity disorder using functional and structural MRI. *IEEE Access*, 5, 23626–23636. DOI: 10.1109/ACCESS.2017.2762703. 32

Zylowska, L., Smalley, S. L., and Schwartz, J. M. (2009). Mindful awareness and ADHD. In F. Didonna (Ed.), *Clinical Handbook of Mindfulness* (pp. 319-338). New York: Springer. DOI: 10.1007/978-0-387-09593-6_18. 93

Authors' Biographies

Dr. Franceli L. Cibrian is an Assistant Professor at the Fowler School of Engineering at Chapman University in Orange, California. She belongs to the Nacional Science System from Mexico, given by Conacyt. She did her postdoctoral training in the STAR lab of Dr. Gillian Hayes at UC Irvine. She received her Ph.D. in Computer Science from the Center for Scientific Research and Higher Education (CICESE) in Mexico, where she worked under the supervision of Dr. Monica Tentori. She has a background in computer system engineering and a minor in software engineering. Her research interests focus on the design, development, and evaluation of ubiquitous interactive technology to support the development of children, particularly children with special needs. She has experience in human–computer interaction and interaction design, with a specific focus on interactive surfaces and educational and therapeutic interventions for children. She has also done research stays at the Design Lab at UC San Diego under the supervision of Dr. Nadir Weibel and in the UCL Interaction Center at University College London (UCL) under the supervision of Dr. Nadia Berthouze.

Dr. Gillian R. Hayes is the Robert A. and Barbara L. Kleist Professor of Informatics at the University of California, Irvine, in the Department of Informatics in the School of Information and Computer Sciences, in the Department of Pediatrics in the School of Medicine, and in the School of Education. Dr. Hayes is also the Vice Provost for Graduate Education and Dean of the Graduate Division at the University of California, Irvine. She is an alumna of Vanderbilt University and the Georgia Institute of Technology. For 15 years, her research has focused on designing, developing, and evaluating technologies in support of vulnerable populations, including those with autism. Building on a background in computer science and a consulting career before academia, she focuses on methods for including people not traditionally represented in the design process or in research. She received a CAREER award from the National Science Foundation in 2008 for her work on mobile technologies for children and families coping with chronic illness and neurodevelopmental disabilities. Her most recent work in this space has focused on wearables and augmented and virtual reality in collaboration with former students and postdoctoral scholars, Dr. LouAnne Boyd, Dr. Franceli Cibrian, Dr. Kathryn Ringland, and Dr. Monica Tentori. Dr. Hayes received the CHI

Social Impact Award in 2019 for her work supporting community-based engaged research related to technologies for neurodevelopmental disorders.

 **Dr. Kimberley D. Lakes** is an Associate Professor of Psychiatry and Neuroscience at the University of California, Riverside. She is an alumna of University of Wisconsin, Madison and completed a clinical child psychology internship at the Children's Hospital Los Angeles and a clinical psychology postdoctoral fellowship at the Children's Hospital of Orange County and University of California, Irvine. She began her academic career with a faculty appointment in the School of Medicine at the University of California, Irvine (UCI), and in 2018, left UCI and joined the faculty of the Department of Psychiatry and Neuroscience at the University of California Riverside (UCR) School of Medicine. Dr. Lakes is a licensed psychologist who practices neuropsychology at the university, with a specialty in child and adolescent neurodevelopmental disorders, particularly ADHD. The central aim of her research is to contribute to scientific understanding of self-regulation and executive functions in children and adolescents. In collaboration with Dr. Hayes and Dr. Cibrian, she received funding from the Agency for Health Research and Quality and National Institutes of Health to develop and study a digital health intervention for children and adolescents with ADHD that includes intervention delivered through Apple Watches for children and a paired iPhone application for parents. She has received a number of awards from the National Institutes of Health for the impact of her work on youth who experience health disparities. Other awards for her work include the Outstanding Recent Graduate Award from the School of Education at the University of Wisconsin, Madison, Aspen Brain Forum Young Investigator Prize (awarded at the international Aspen Brain Forum hosted by the New York Academy of Sciences, Aspen Brain Foundation, and National Science Foundation), and a research fellowship to the University of Oxford in the United Kingdom. www.drkimberleylakes.com.

Printed in the United States
by Baker & Taylor Publisher Services